Dimensions Math®
Textbook 1A

Authors and Reviewers

Bill Jackson

Jenny Kempe

Cassandra Turner

Allison Coates

Tricia Salerno

Pearly Yuen

Consultant

Dr. Richard Askey

Singapore Math Inc.

Published by Singapore Math Inc.

19535 SW 129th Avenue
Tualatin, OR 97062
www.singaporemath.com

Dimensions Math® Textbook 1A
ISBN 978-1-947226-04-3

First published 2018
Reprinted 2019 (twice), 2020 (twice), 2021

Printed in China

Acknowledgments

Editing by the Singapore Math Inc. team.
Design and illustration by Cameron Wray.

Preface

The Dimensions Math® Pre-Kindergarten to Grade 5 series is based on the pedagogy and methodology of math education in Singapore. The curriculum develops concepts in increasing levels of abstraction, emphasizing the three pedagogical stages: Concrete, Pictorial, and Abstract. Each topic is introduced, then thoughtfully developed through the use of problem solving, student discourse, and opportunities for mastery of skills.

Features and Lesson Components

Students work through the lessons with the help of five friends: Emma, Alex, Sofia, Dion, and Mei. The characters appear throughout the series and help students develop metacognitive reasoning through questions, hints, and ideas.

The colored boxes and blank lines in the textbook lessons are used to facilitate student discussion. Rather than writing in the textbooks, students can use whiteboards or notebooks to record their ideas, methods, and solutions.

Chapter Opener

Each chapter begins with an engaging scenario that stimulates student curiosity in new concepts. This scenario also provides teachers an opportunity to review skills.

Think

Students, with guidance from teachers, solve a problem using a variety of methods.

Learn

One or more solutions to the problem in **Think** are presented, along with definitions and other information to consolidate the concepts introduced in **Think**.

Do

A variety of practice problems allow teachers to lead discussion or encourage independent mastery. These activities solidify and deepen student understanding of the concepts.

Exercise

A pencil icon ━━━━━━▶ at the end of the lesson links to additional practice problems in the workbook.

Practice

Periodic practice provides teachers with opportunities for consolidation, remediation, and assessment.

Review

Cumulative reviews provide ongoing practice of concepts and skills.

Emma Alex Sofia Dion Mei

Contents

Chapter	Lesson	Page

Chapter		Lesson	Page

Chapter	Lesson	Page

Chapter 1

Numbers to 10

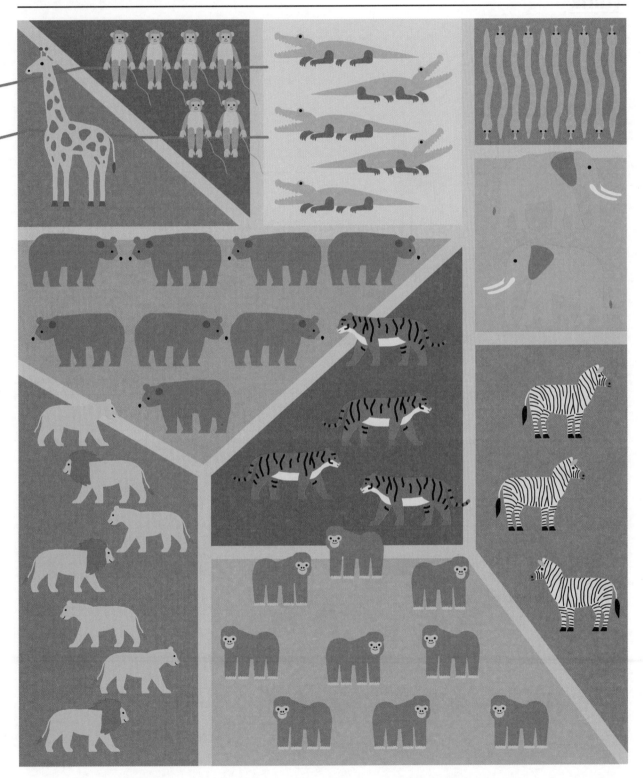

Think

How many?

Learn

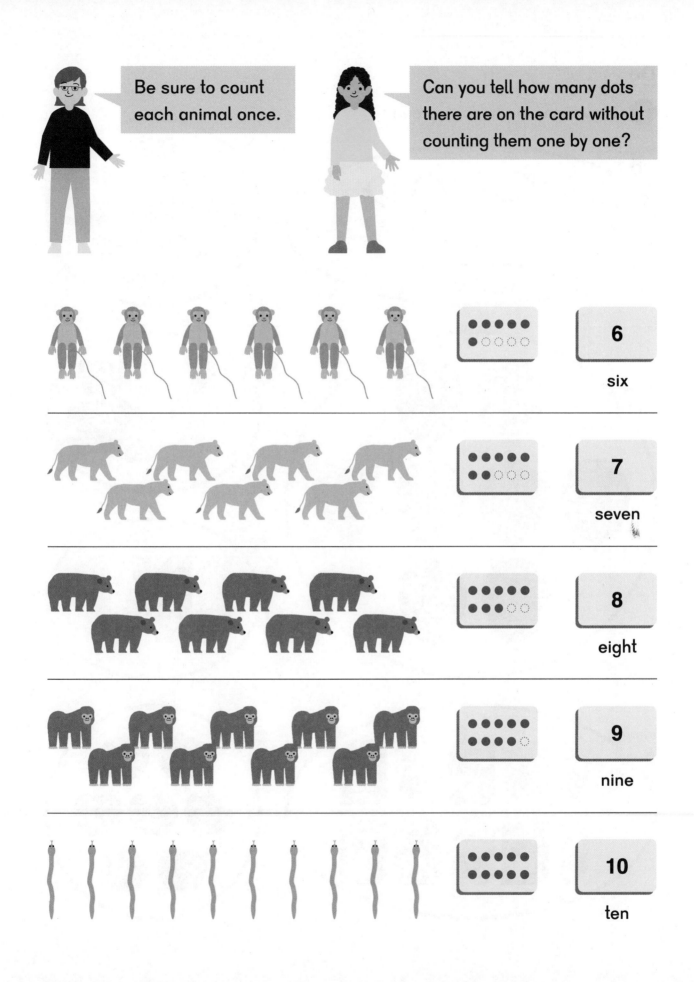

Do

1 How many are in each group?

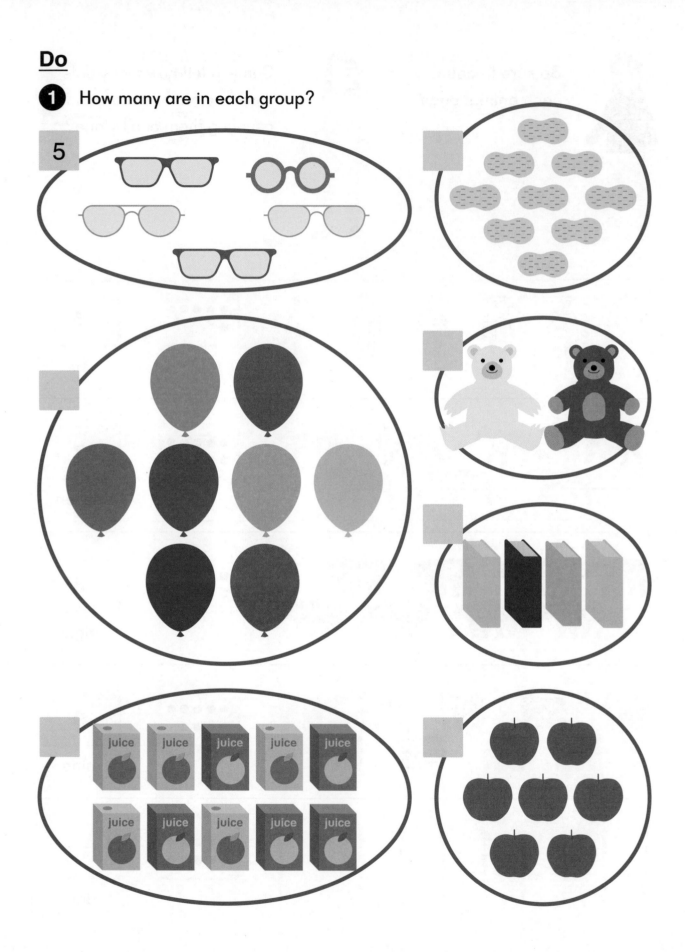

5

2 How many ● are there?

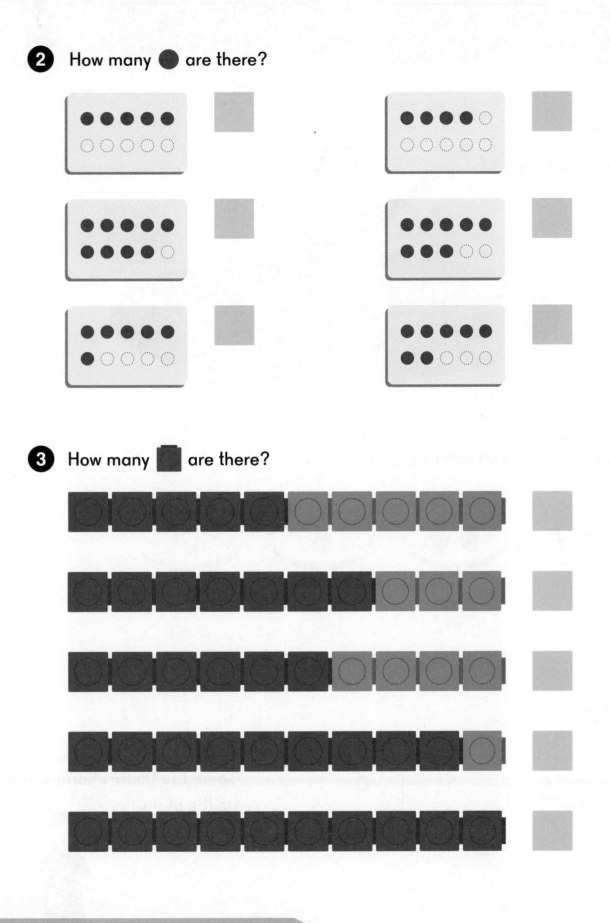

3 How many ▪ are there?

Think

Alex is eating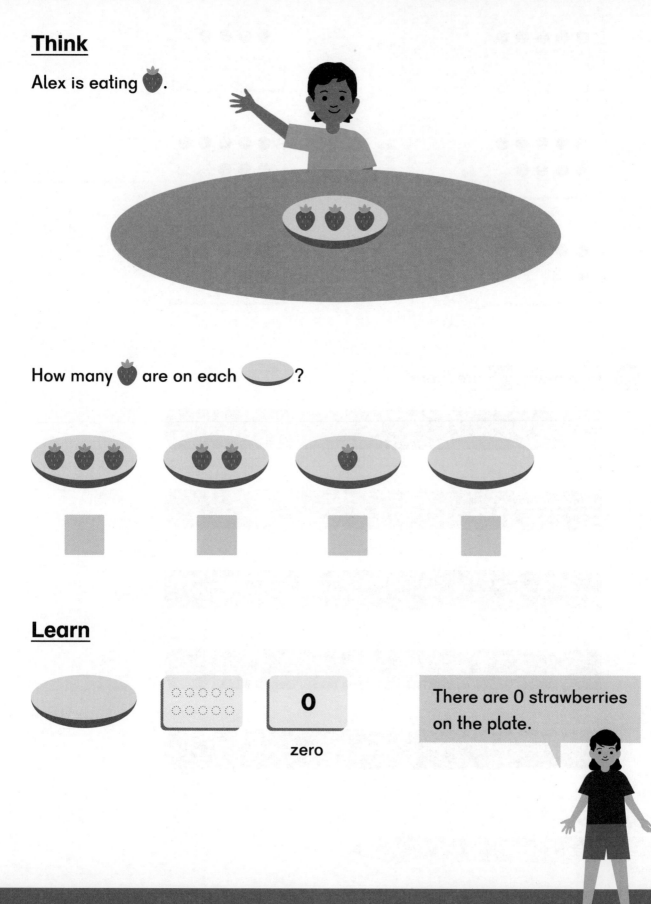

How many 🍓 are on each 🍽 ?

Learn

0

zero

There are 0 strawberries on the plate.

Do

1 Count back from 10 to 0.

2 How many 🔴 are in each 🥣?

[   ]   [   ]   [   ]

3 How many are jumping on the 🛏️ ?

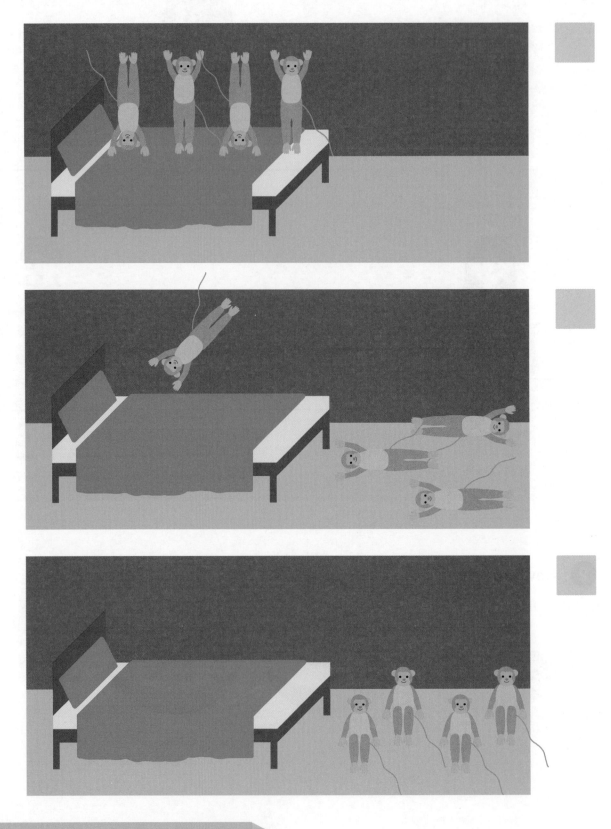

Exercise 2 • page 3

1-2 The Number 0

Think

Put the numbers in order from least to greatest.

2　　10　　1　　8　　3

5　　7　　4　　6　　0　　9

Learn

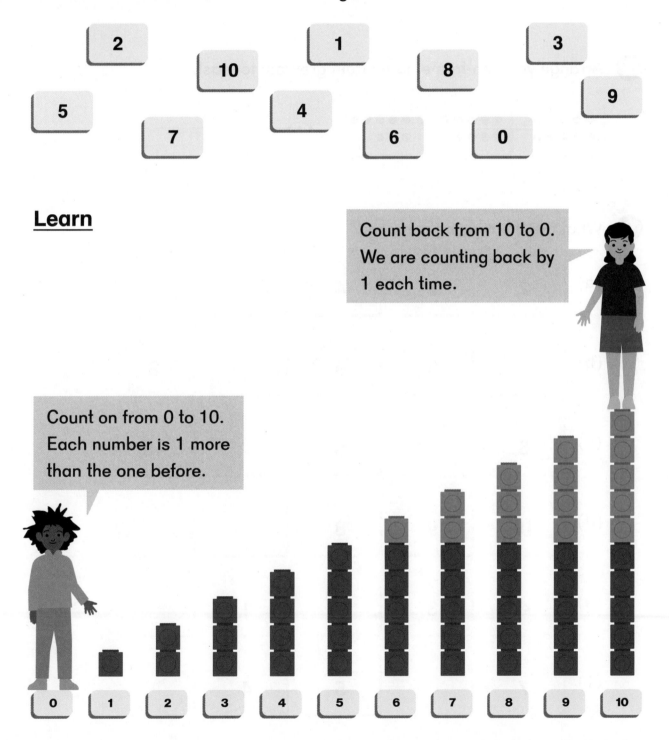

Count back from 10 to 0. We are counting back by 1 each time.

Count on from 0 to 10. Each number is 1 more than the one before.

0　1　2　3　4　5　6　7　8　9　10

Do

1 Arrange your ten-frame cards from least to greatest.

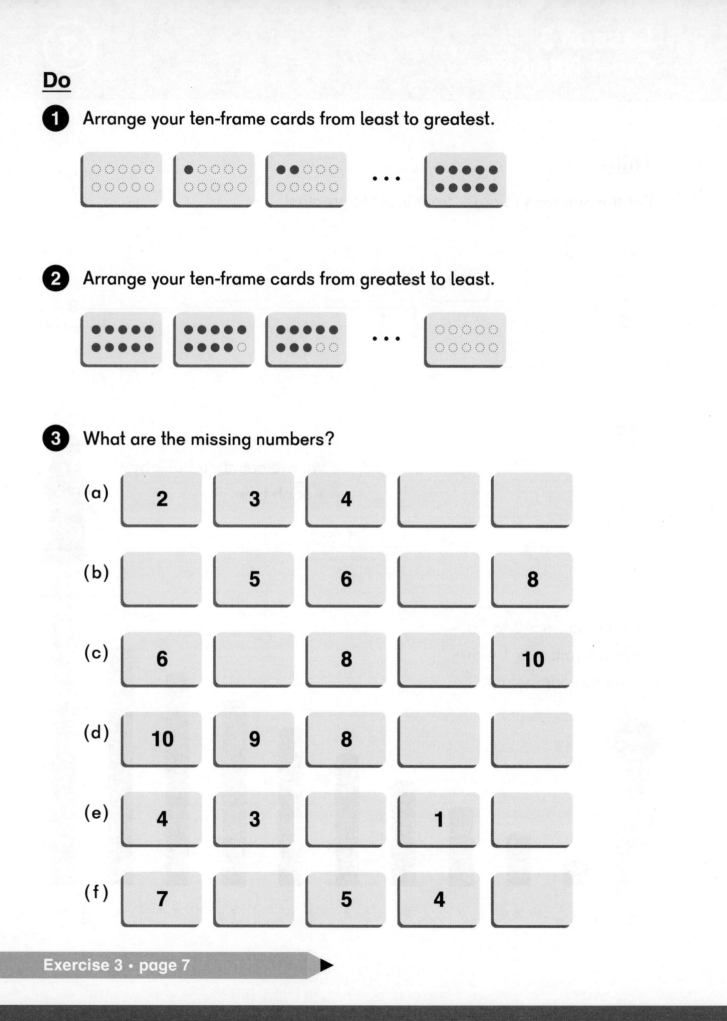

2 Arrange your ten-frame cards from greatest to least.

3 What are the missing numbers?

(a) | 2 | 3 | 4 | | |

(b) | | 5 | 6 | | 8 |

(c) | 6 | | 8 | | 10 |

(d) | 10 | 9 | 8 | | |

(e) | 4 | 3 | | 1 | |

(f) | 7 | | 5 | 4 | |

Think

Are there **more** dogs or cats?

Are there **fewer** dogs or cats?

Learn

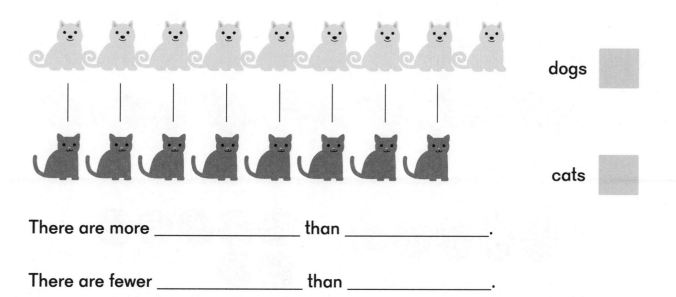

dogs

cats

There are more _____ than _____.

There are fewer _____ than _____.

Do

1 Which has more?

(a)

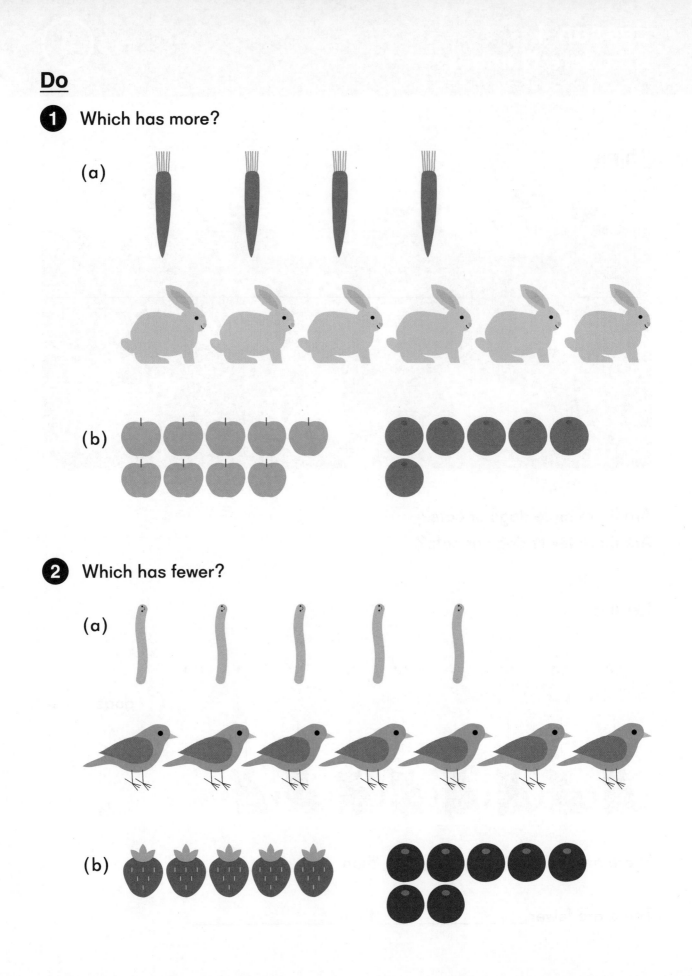

(b)

2 Which has fewer?

(a)

(b)

3 More or fewer?

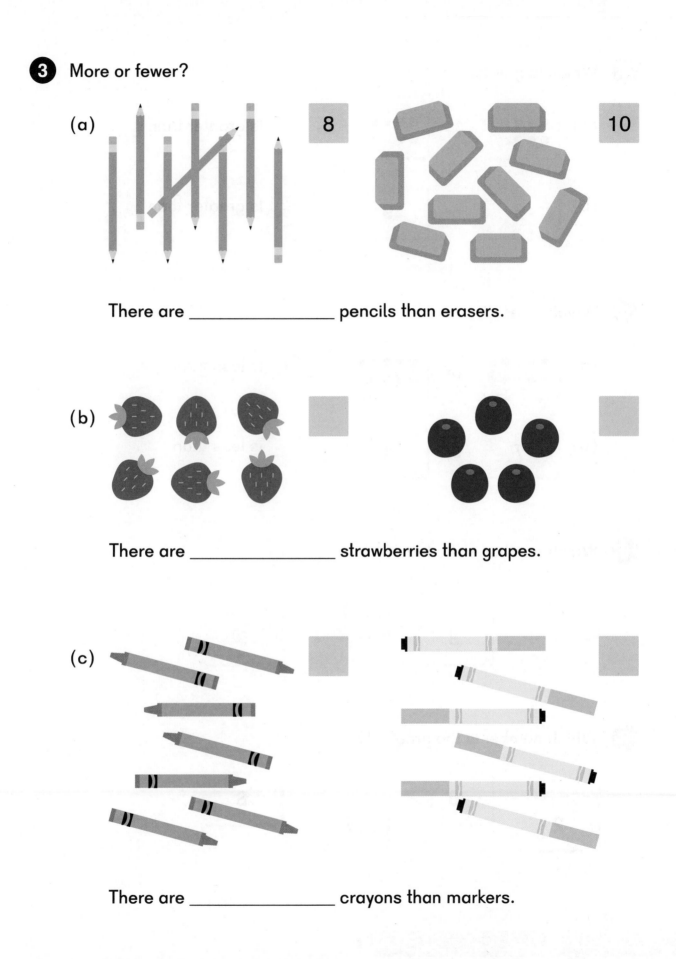

(a) | 8 | | 10 |

There are _____ pencils than erasers.

(b) | | | |

There are _____ strawberries than grapes.

(c) | | | |

There are _____ crayons than markers.

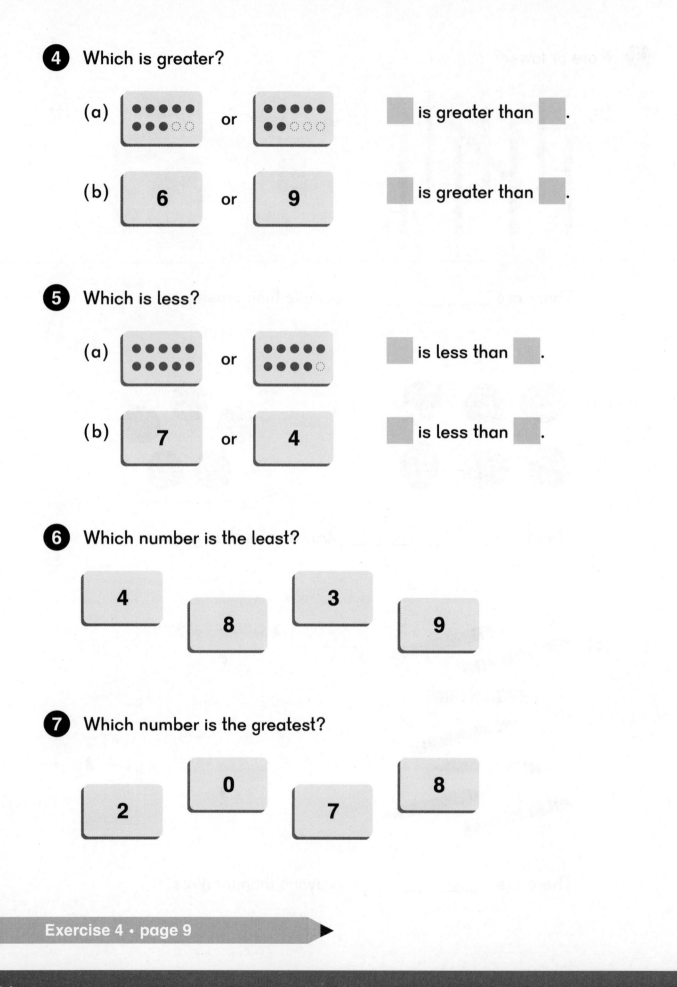

4 Which is greater?

(a) [●●●●● / ●●●●○○] or [●●●●● / ●●○○○]   ⬜ is greater than ⬜.

(b) 6 or 9   ⬜ is greater than ⬜.

5 Which is less?

(a) [●●●●● / ●●●●●] or [●●●●● / ●●●●○]   ⬜ is less than ⬜.

(b) 7 or 4   ⬜ is less than ⬜.

6 Which number is the least?

4   8   3   9

7 Which number is the greatest?

2   0   7   8

Exercise 4 • page 9

1 How many ● are there?

(a)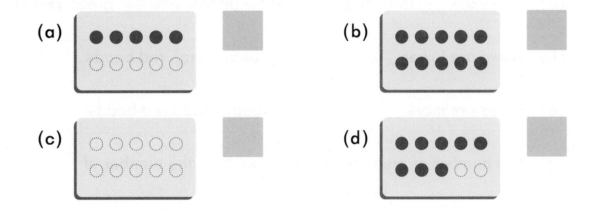

(b)

(c)

(d)

2 What are the missing numbers?

(a) | 5 | 6 | | | 9 |

(b) | 4 | | 2 | 1 | |

3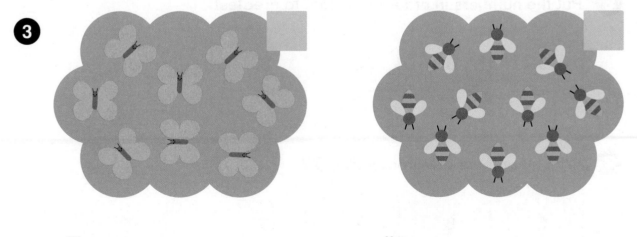

There are more _____ than _____.

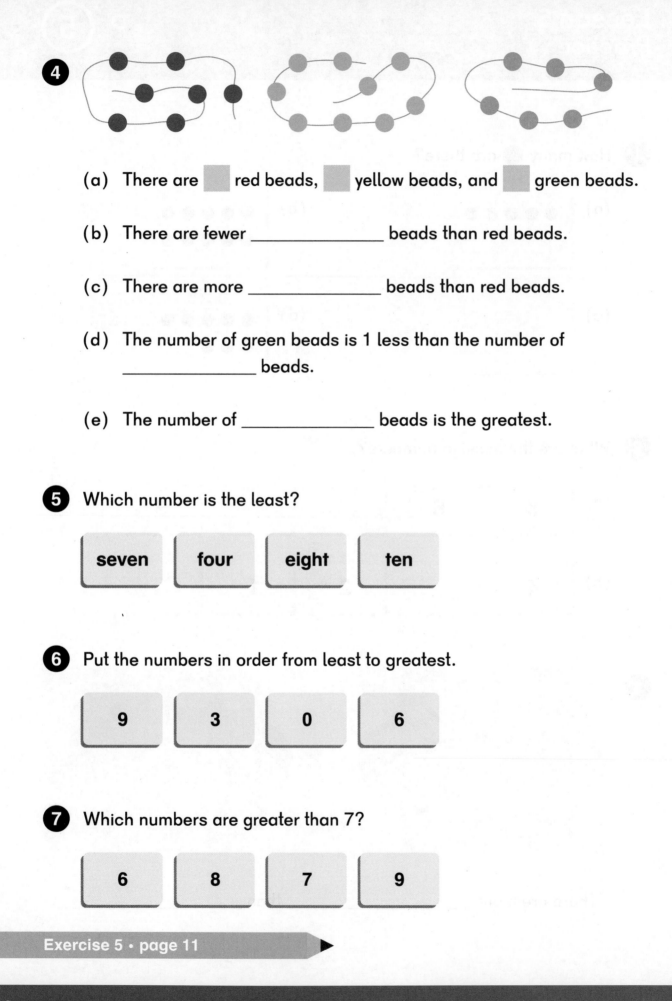

4

(a) There are ☐ red beads, ☐ yellow beads, and ☐ green beads.

(b) There are fewer _____ beads than red beads.

(c) There are more _____ beads than red beads.

(d) The number of green beads is 1 less than the number of _____ beads.

(e) The number of _____ beads is the greatest.

5 Which number is the least?

| seven | four | eight | ten |

6 Put the numbers in order from least to greatest.

| 9 | 3 | 0 | 6 |

7 Which numbers are greater than 7?

| 6 | 8 | 7 | 9 |

Exercise 5 • page 11

Chapter 2

Number Bonds

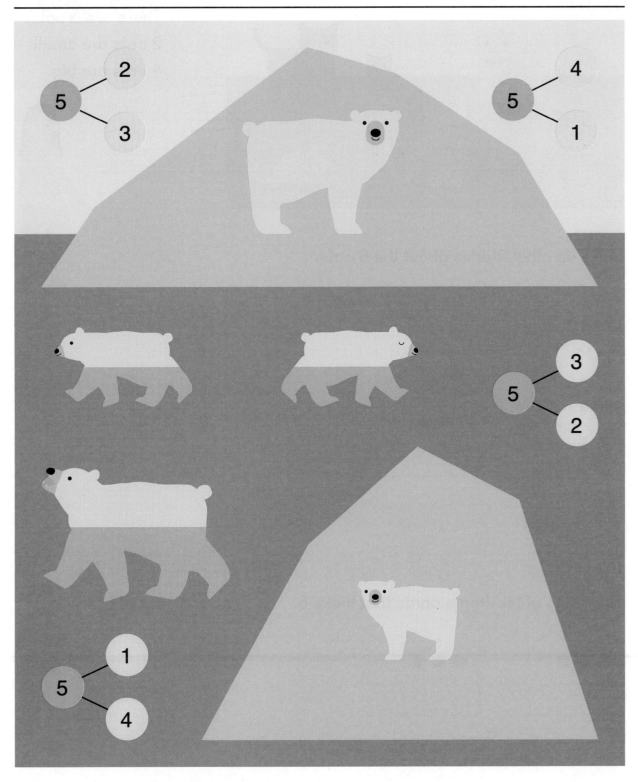

Think

There are 6 cats.
2 cats are small.
4 cats are big.

Make up other stories about the 6 cats.

Learn

part

whole

6

2

4

part

6 is [ ] and [ ].

Find pairs of ten-frame cards that make 6.

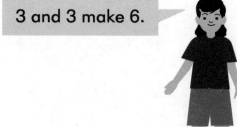

3 and 3 make 6.

Do

1 How many more ● make 6?

(a)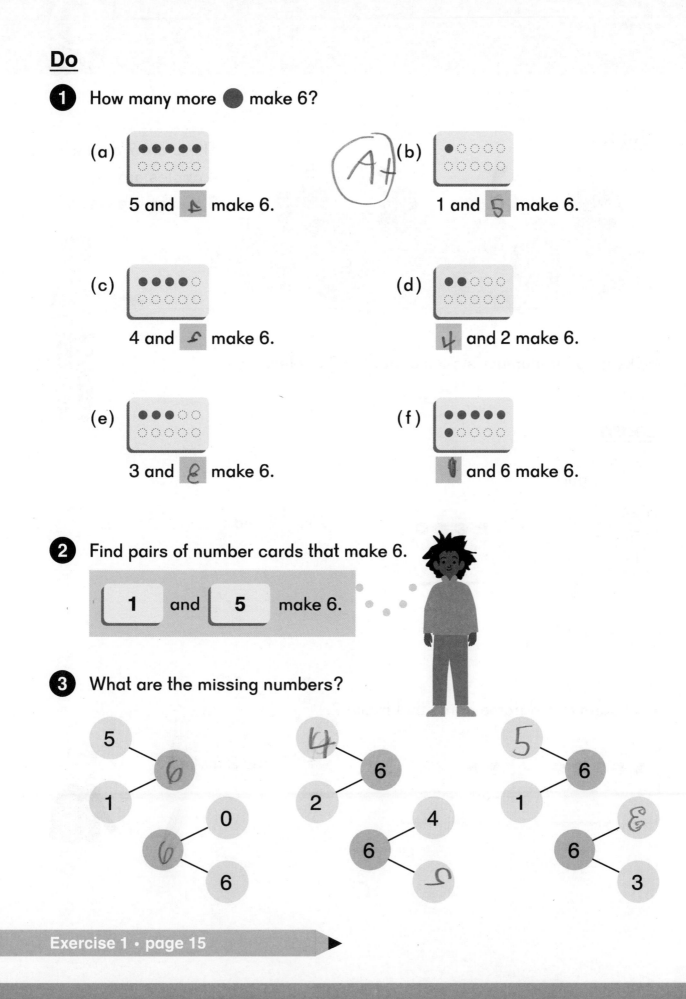
5 and 1 make 6.

(b)
1 and 5 make 6.

(A+)

(c)
4 and 2 make 6.

(d)
4 and 2 make 6.

(e)
3 and 3 make 6.

(f)
0 and 6 make 6.

2 Find pairs of number cards that make 6.

1 and 5 make 6.

3 What are the missing numbers?

5
1   6

6   0
      6

4
2   6

6   4
      2

5
1   6

6   3
      3

Think

There are 7 apples.
4 apples are red.
3 apples are green.

Make up other number stories about the 7 apples.

Learn

part

whole

7

4

3

part

7 is ⬜ and ⬜.

Find pairs of ten-frame cards that make 7.

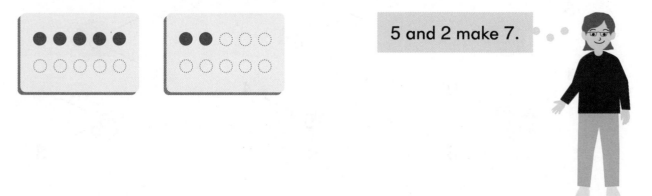

5 and 2 make 7.

Do

1 How many more ⬤ make 7?

(a) 6 and ☐ make 7.

(b) 1 and ☐ make 7.

(c) 5 and ☐ make 7.

(d) ☐ and 2 make 7.

(e) ☐ and 4 make 7.

(f) 3 and ☐ make 7.

(g) 7 and ☐ make 7.

(h) 0 and ☐ make 7.

2 What are the missing numbers?

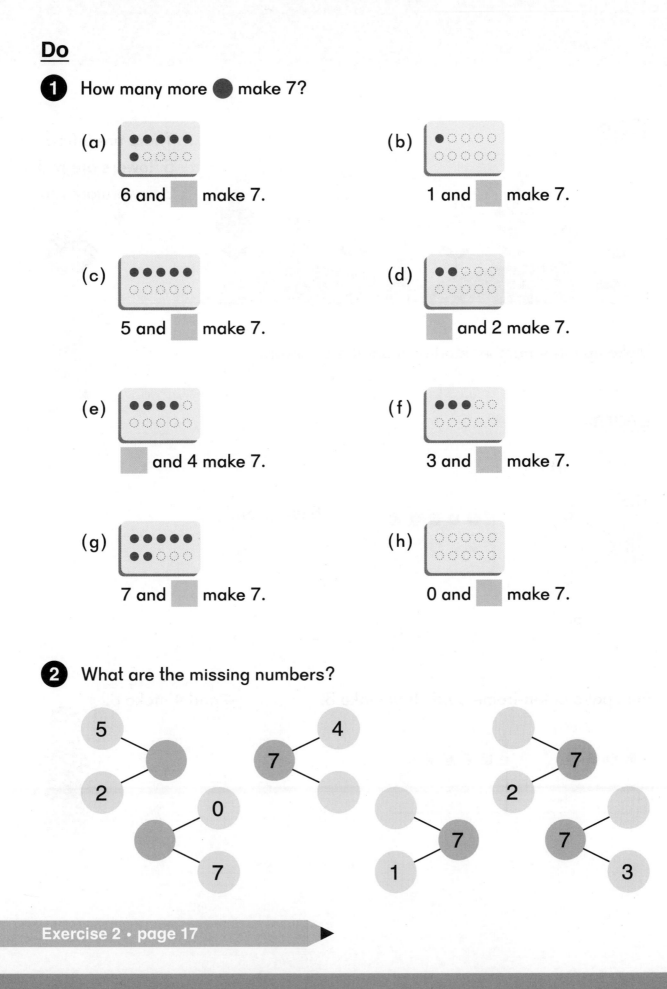

Exercise 2 • page 17

Think

There are 8 flowers.
5 flowers are pink.
3 flowers are red.

Make up other number stories about the 8 flowers.

Learn

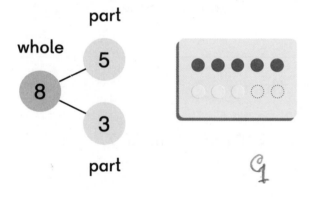

part

whole

8 is [4] and [4].

Find pairs of ten-frame cards that make 8.

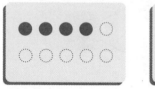
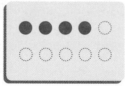

4 and 4 make 8.

Do

1 How many more ⬤ make 8?

(a)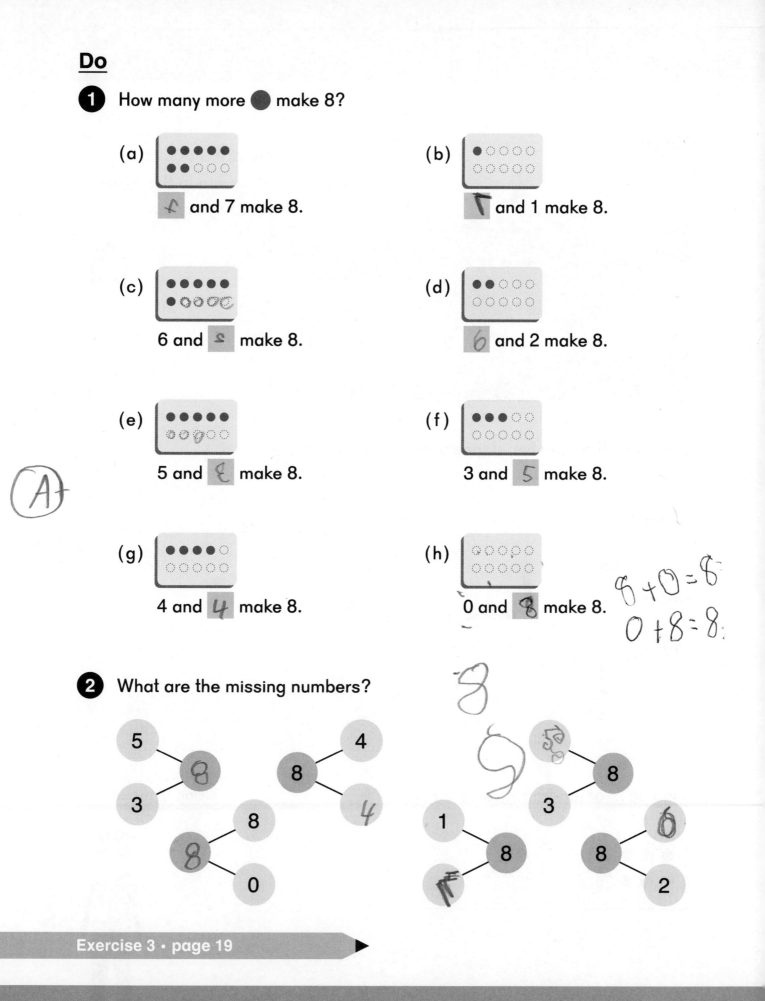

_1_ and 7 make 8.

(b)

_7_ and 1 make 8.

(c)

6 and _2_ make 8.

(d)

_6_ and 2 make 8.

(e)

5 and _3_ make 8.

(f)

3 and _5_ make 8.

(g)

4 and _4_ make 8.

(h)

0 and _8_ make 8.

8 + 0 = 8
0 + 8 = 8

2 What are the missing numbers?

Think

There are 9 ducks.
5 ducks are in the water.
4 ducks are on land.

Make up other number stories about the 9 ducks.

Learn

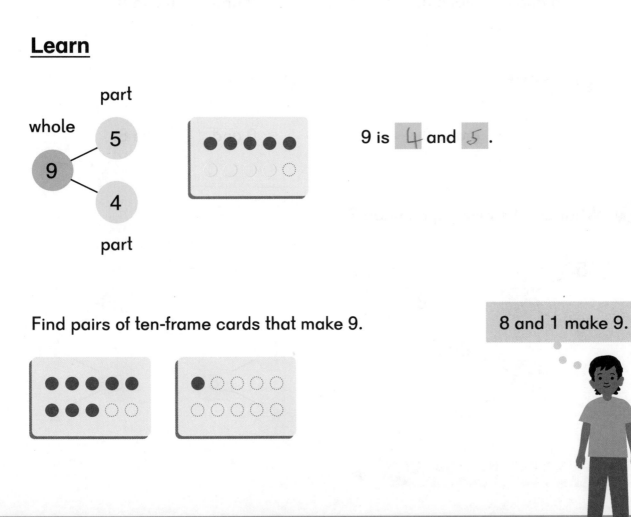

part

whole

5

9

4

part

9 is 4 and 5 .

Find pairs of ten-frame cards that make 9.

8 and 1 make 9.

Do

1 How many more ● make 9?

(a)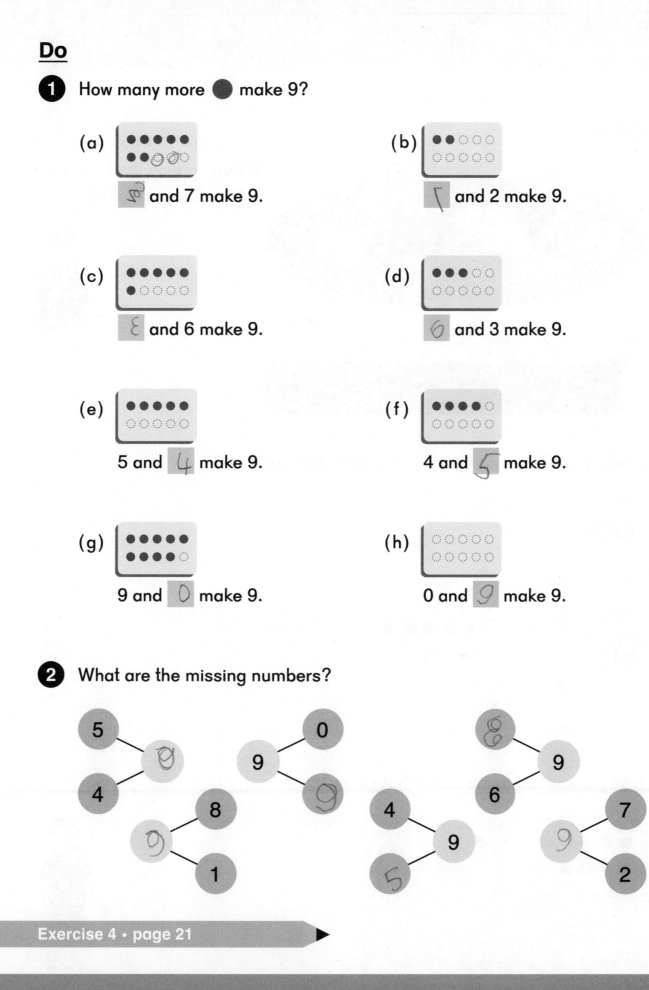

2 and 7 make 9.

(b)

7 and 2 make 9.

(c)

3 and 6 make 9.

(d)

6 and 3 make 9.

(e)

5 and *4* make 9.

(f)

4 and *5* make 9.

(g)

9 and *0* make 9.

(h)

0 and *9* make 9.

2 What are the missing numbers?

5
4
9

9
8
1

0
9
9

4
9
5

8
6
9

9
7
2

Think

There are 10 frogs.
6 frogs are on the lily pad.
4 frogs are on the log.

Make up other number stories about the 10 frogs.

Learn

part

whole

6

10

4

part

10 is ⬚5⬚ and ⬚5⬚.

Find pairs of ten-frame cards that make 10.

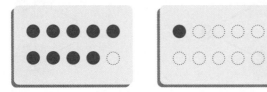

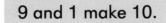
9 and 1 make 10.

Do

1 How many more ● make 10?

(a)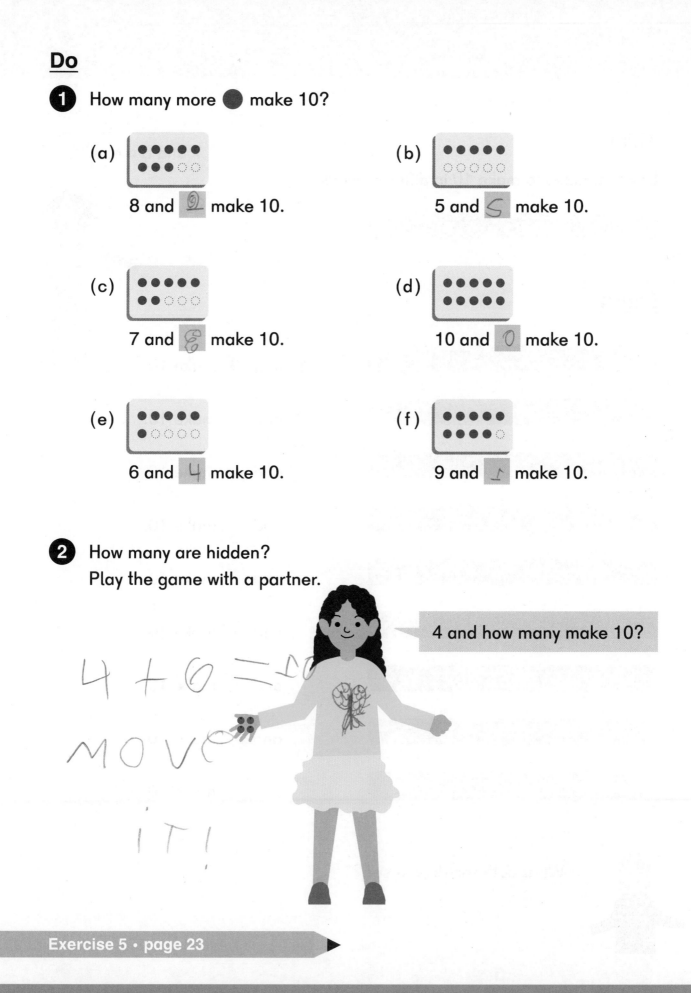

8 and 2 make 10.

(b)

5 and 5 make 10.

(c)

7 and 3 make 10.

(d)

10 and 0 make 10.

(e)

6 and 4 make 10.

(f)

9 and 1 make 10.

2 How many are hidden?
Play the game with a partner.

4 and how many make 10?

4 + 6 = 10

MOVE

IT!

Exercise 5 • page 23

Think

Use the cubes to make 10 in different ways.

7 and 3 make 10.

Learn

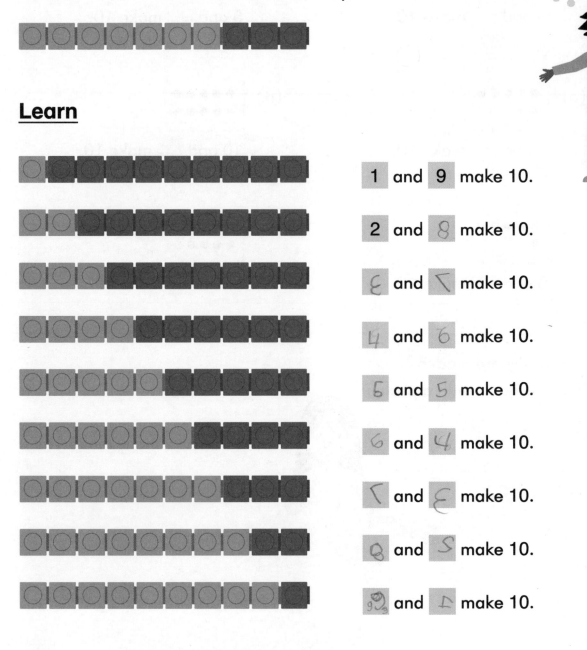

1 and 9 make 10.

2 and 8 make 10.

3 and 7 make 10.

4 and 6 make 10.

5 and 5 make 10.

6 and 4 make 10.

7 and 3 make 10.

8 and 2 make 10.

9 and 1 make 10.

What patterns do you see?

Do

1 Line up 2 sets of number cards so that
the top card makes 10 with the one under it.

| 0 | 1 | 2 | 3 | 4 | 5 | 6 | 7 | 8 | 9 | 10 |

| 10 | 9 | 8 | 7 | 6 | 5 | 4 | 3 | 2 | 1 | 0 |

2 What are the missing numbers?

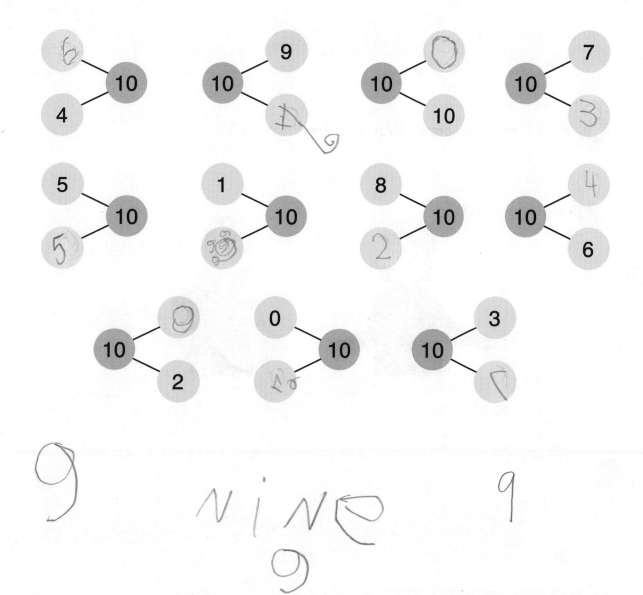

Roll a die.

Say the number that makes a 10 with the number on the die.

2 and 8 make 10.

4 Find 2 number cards that together make 10.

7 and 3 make 10.

10
2 ☐

Exercise 6 • page 25

2-6 Make 10 — Part 2

1

How many more does Emma need to have 10?

4

2 How many more stars are needed to complete the number bonds?

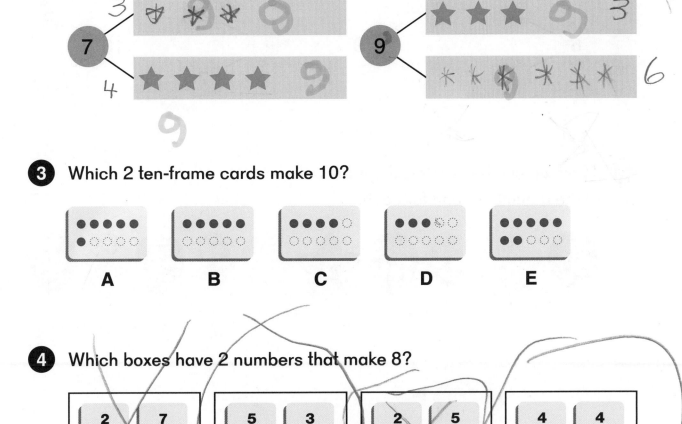

3

7

4

9

3

6

3 Which 2 ten-frame cards make 10?

A   B   C   D   E

4 Which boxes have 2 numbers that make 8?

| 2 | 7 | | 5 | 3 | | 2 | 5 | | 4 | 4 |

A   B   C   D

Find pairs of numbers in each box that make 9.

5 Find pairs of numbers in each box that make 9.

(a)

3	2	4
8	5	6

(b)

9	8	2
4	7	0

6 Complete the number bonds.

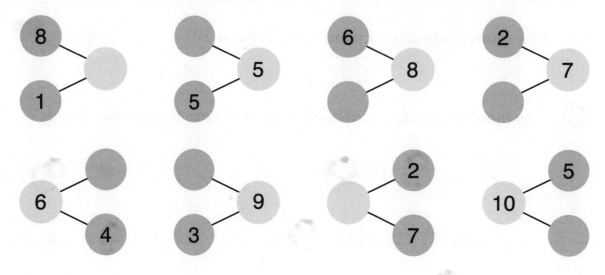

7 Make up number stories about the balloons.
Write a number bond for each.

Exercise 7 • page 27

2-7 Practice

Chapter 3

Addition BY BANANA!

There are 3 blue mugs.
There are 2 red mugs.
There are 5 mugs altogether.

What other number stories can you make?

Think

How many birds will there be on the branch altogether?

Learn

5 + 3 = 8

part

5

whole

?

3

part

There will be 8 birds on the branch altogether.

This is **addition**.
+ is read as plus.

= is read as equals.
It means **the same value as.**
5 + 3 = 8 is an **equation**.

Do

1

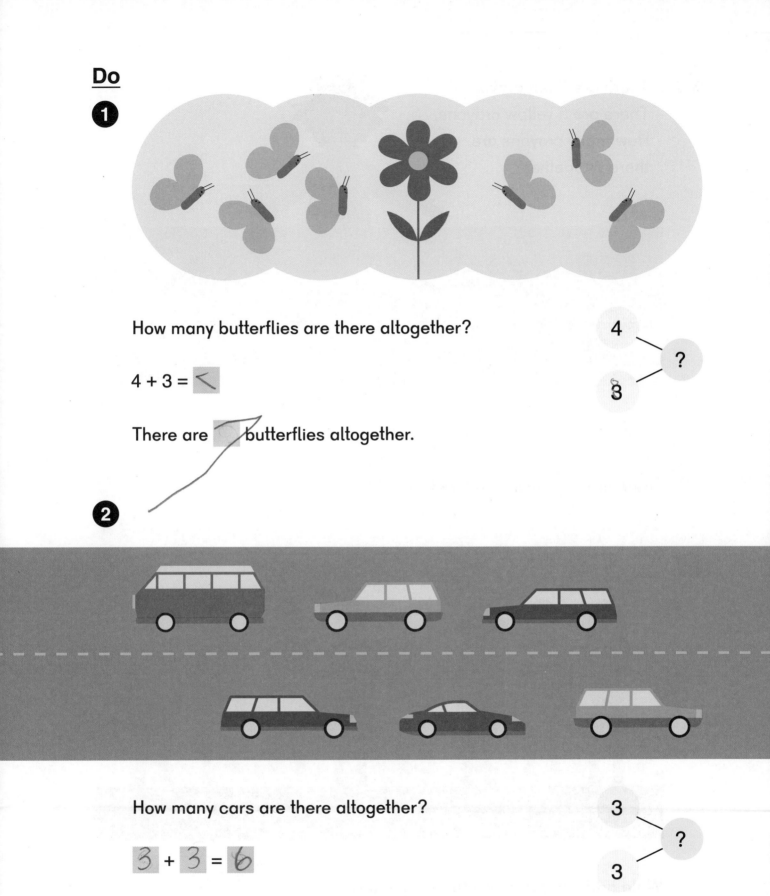

How many butterflies are there altogether?

4 + 3 =

There are ☐ butterflies altogether.

4
3
?

2

How many cars are there altogether?

3 + 3 = 6

There are ☐ cars altogether.

3
3
?

3 There are 5 red crayons.
There are 5 yellow crayons.
How many crayons are
there altogether?

 + 6 = 10

There are 10 crayons altogether.

4

Write addition equations for this picture. 4 + 4 = 8

Exercise 1 • page 31

3-1 Addition as Putting Together

Think

5 birds are on a branch.

2 more birds come.

How many birds will be on the branch altogether?

(handwritten)
3+2 = 5
5 + 5 = 10
10 + 10 = 20
20 + 20 = 40
40 + 40 = 80
80 + 80 =

5
?
2

100 & 60

Learn

5 + 2 = 7

There will be 7 birds on the branch altogether.

Do

1

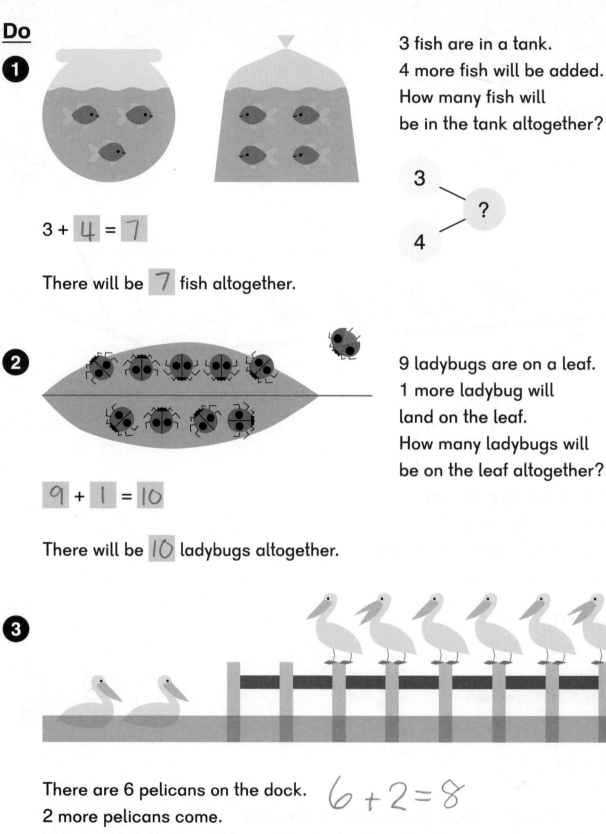

3 fish are in a tank.
4 more fish will be added.
How many fish will
be in the tank altogether?

3

4

?

3 + 4 = 7

There will be 7 fish altogether.

2

9 ladybugs are on a leaf.
1 more ladybug will
land on the leaf.
How many ladybugs will
be on the leaf altogether?

9 + 1 = 10

There will be 10 ladybugs altogether.

3

There are 6 pelicans on the dock.   6 + 2 = 8
2 more pelicans come.
How many pelicans will be on the dock altogether?
Write an addition equation.

Exercise 2 • page 33

Think

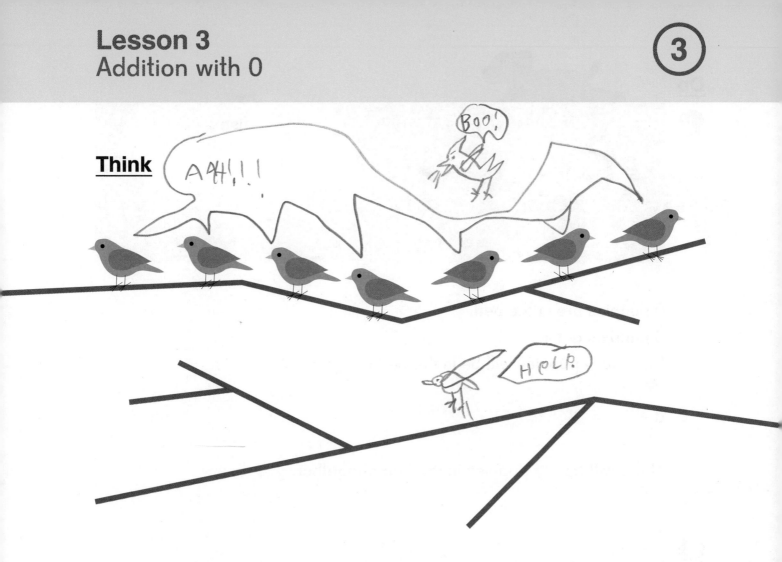

There are 7 birds on one branch.

There are 0 birds on the other branch.

Write an addition equation to show how many birds there are in all.

Learn

7 + 0 = 7

There are 7 birds in all.

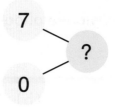

1

0 puppies are in the bed.
3 puppies get in.
How many puppies will be in the bed altogether?

0 + 3 = 3

There will be 3 puppies in the bed altogether.

2

How many sandwiches are there altogether?

0 + 0 = 0

There are 0 sandwiches altogether.

3 (a) 7 + 0 = 7   (b) 0 + 6 = 6   (c) 8 = 8 + 0

    (d) 0 + 2 = 2   (e) 9 + 0 = 9   (f) 10 = 0 + 10

Exercise 3 • page 35

Think

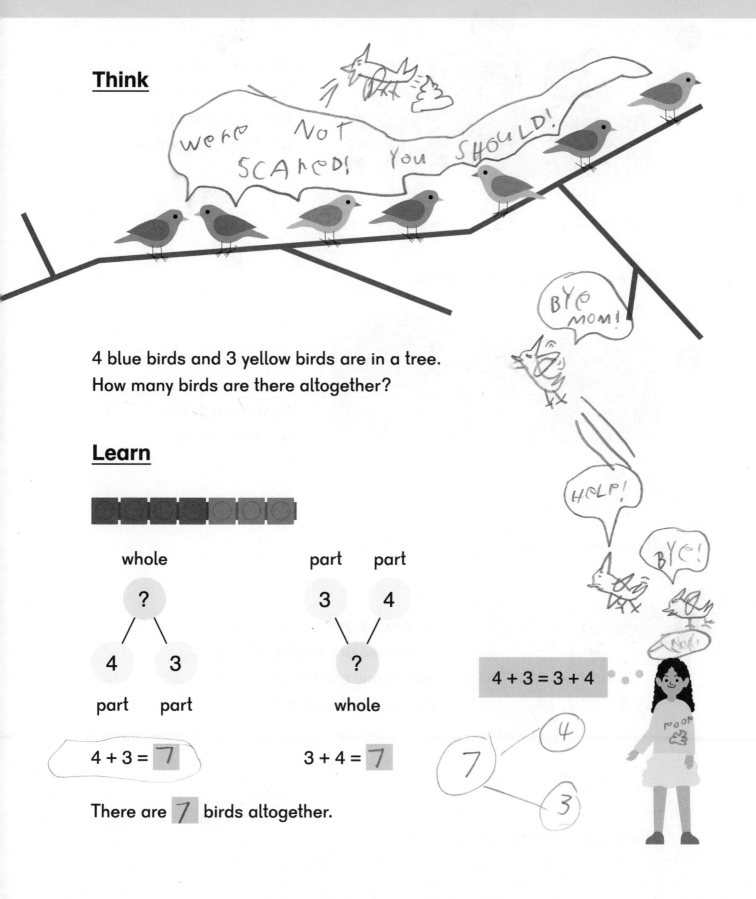

4 blue birds and 3 yellow birds are in a tree.
How many birds are there altogether?

Learn

whole

part   part

?

3   4

4   3

?

part   part

whole

4 + 3 = 7

3 + 4 = 7

There are 7 birds altogether.

4 + 3 = 3 + 4

7

4

3

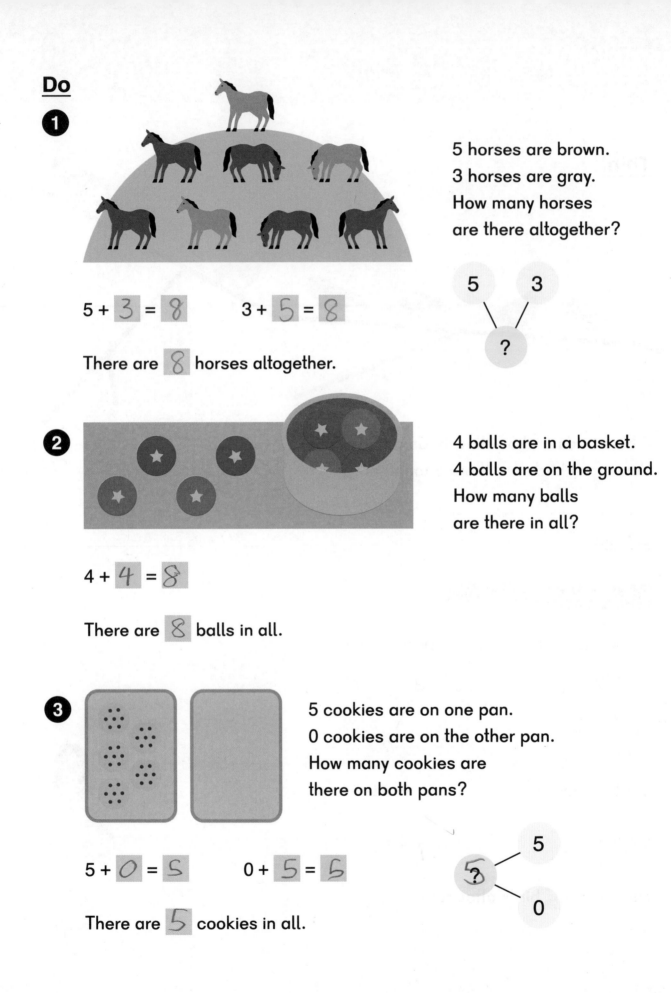

5 horses are brown.
3 horses are gray.
How many horses
are there altogether?

5 + 3 = 8   3 + 5 = 8

There are 8 horses altogether.

2

4 balls are in a basket.
4 balls are on the ground.
How many balls
are there in all?

4 + 4 = 8

There are 8 balls in all.

3

5 cookies are on one pan.
0 cookies are on the other pan.
How many cookies are
there on both pans?

5 + 0 = 5   0 + 5 = 5

There are 5 cookies in all.

4 (a)

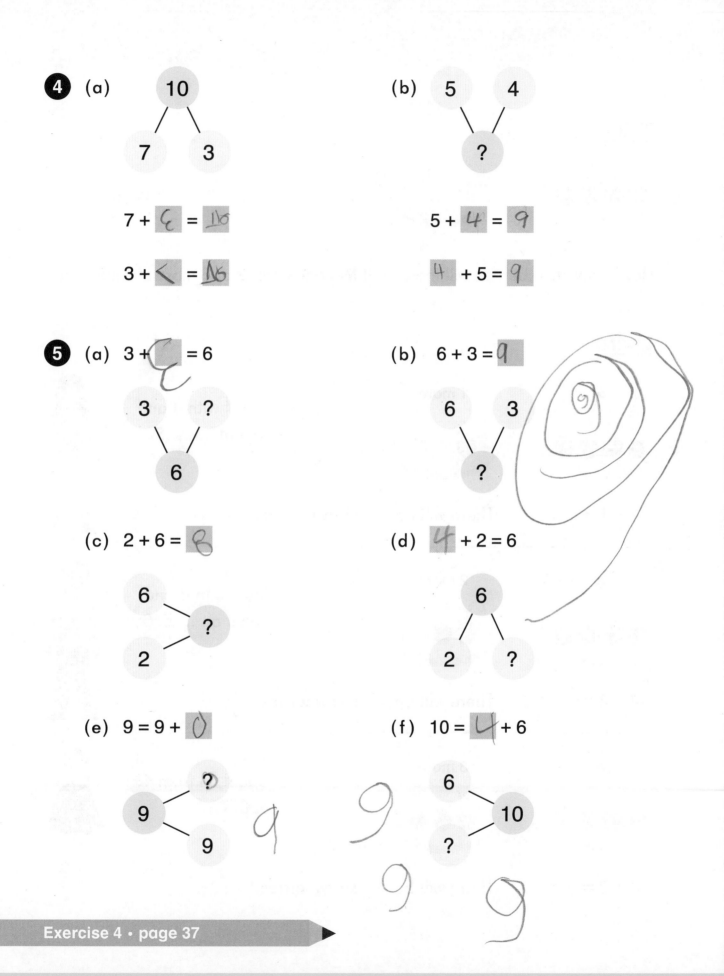

10

7   3

7 + 𝟼 = 10

3 + ＜ = 1̶0̶

(b)

5   4

?

5 + 4 = 9

4 + 5 = 9

5 (a) 3 + 𝟼 = 6

3   ?

6

(b) 6 + 3 = 9

6   3

?

(c) 2 + 6 = 𝟾

6

2   ?

(d) 4 + 2 = 6

6

2   ?

(e) 9 = 9 + 0

?

9

9

(f) 10 = 4 + 6

6

10

?

Exercise 4 • page 37

Think

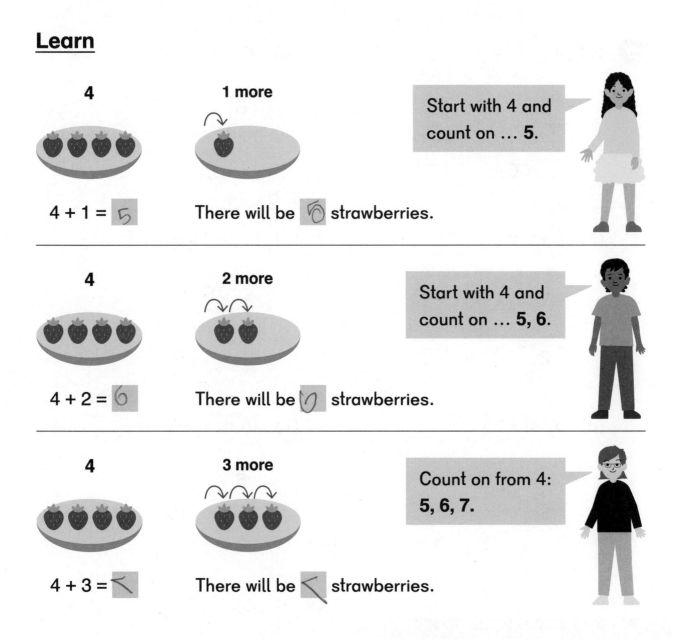

How many strawberries will there be if Mei adds 1 or 2 or 3 more strawberries?

Learn

4   **1 more**

Start with 4 and count on ... **5.**

$4 + 1 = \boxed{5}$ There will be $\boxed{6}$ strawberries.

---

4   **2 more**

Start with 4 and count on ... **5, 6.**

$4 + 2 = \boxed{6}$ There will be $\boxed{6}$ strawberries.

---

4   **3 more**

Count on from 4: **5, 6, 7.**

$4 + 3 = \boxed{7}$ There will be $\boxed{7}$ strawberries.

Do

1 Add 6 and 2.

Count on from 6: **7, 8**.

| 3 | 4 | 5 | 6 | 7 | 8 | 9 |

$6 + 2 = 8$

2 Add 3 and 6.

Count on from 6: **7, 8, 9**.

$3 + 6 = 6 + 3$.

| 3 | 4 | 5 | 6 | 7 | 8 | 9 | 10 |

$3 + 6 = 9$

3 (a) $5 + 1 = 6$ (b) $1 + 6 = 7$ (c) $9 = 1 + 8$

     (d) $4 + 2 = 6$ (e) $2 + 7 = 9$ (f) $10 = 8 + 2$

     (g) $5 + 3 = 8$ (h) $3 + 4 = 7$ (i) $9 = 6 + 3$

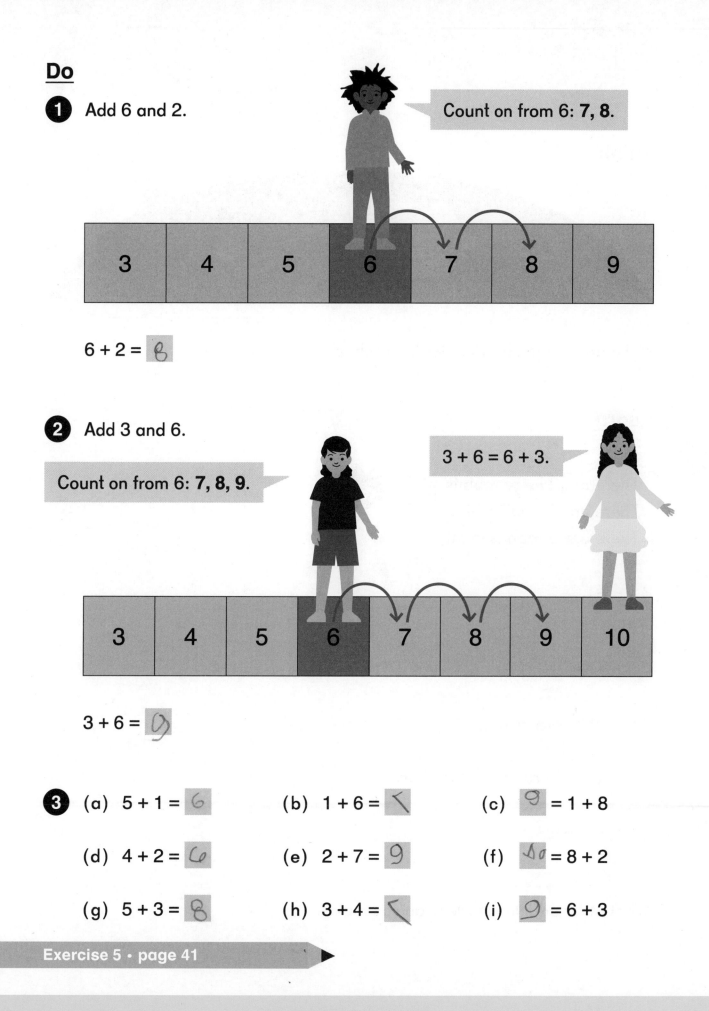

Exercise 5 • page 41

Lesson 6
Make Addition Stories

Think

Make up an addition story for the rabbits.

Learn

There are 2 large rabbits.
There are 4 small rabbits.
There are 6 rabbits in all.

2 + 4 = 6

2
4
6

3 rabbits are gray.

3 rabbits are brown.

There are 6 rabbits altogether.

? ?
?

 +  =

1 + 5 = 6
0 + 6 = 6

What other addition stories can you make?

Do

1 Make up addition stories for each equation.

(a)

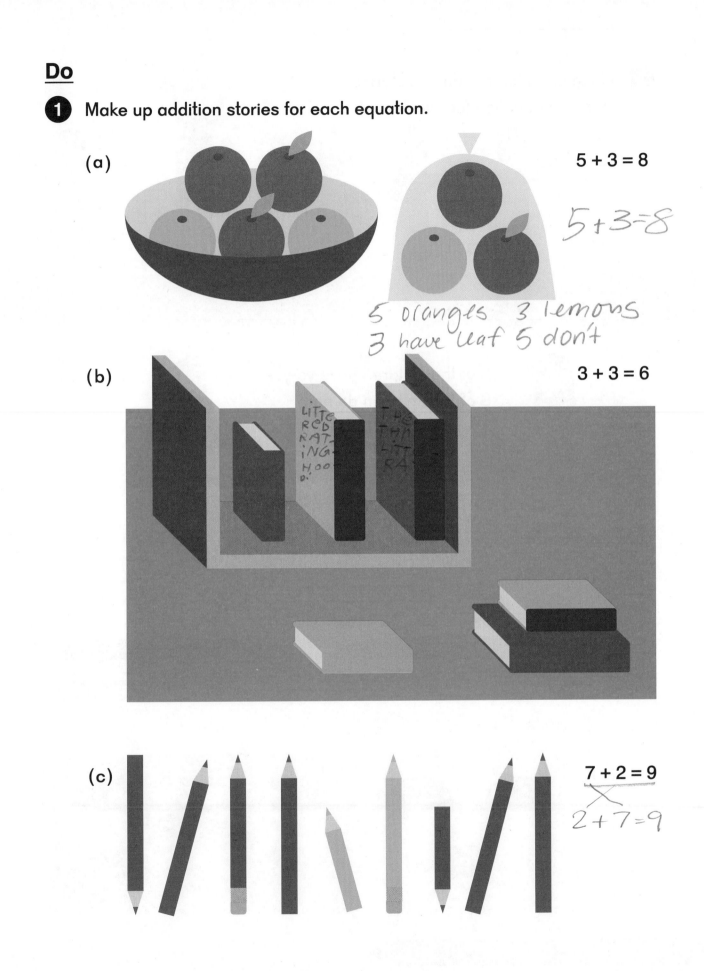

5 + 3 = 8

5+3=8

5 oranges   3 lemons
3 have leaf 5 don't

(b)

3 + 3 = 6

(c)

7 + 2 = 9

2+7=9

2 Make up different addition stories.
Write an equation for each story.

(a)

2 of the children have yo-yos.
3 of the children do not.
There are 5 children in all.

2 + 3 = 5

3 + 2 = 5

(b)

3 4 5 6

0

3 red + 4 purple = 7

(c)

1 + 7 = 8   2 + 6 = 8
7 + 1 = 8   3 + 5 = 8

Exercise 6 • page 43

3-6 Make Addition Stories

Think   Say the answers and look for patterns.

1 + 1	2 + 1	3 + 1	4 + 1	5 + 1	6 + 1	7 + 1	8 + 1	9 + 1
1 + 2	2 + 2	3 + 2	4 + 2	5 + 2	6 + 2	7 + 2	8 + 2	
1 + 3	2 + 3	3 + 3	4 + 3	5 + 3	6 + 3	7 + 3		
1 + 4	2 + 4	3 + 4	4 + 4	5 + 4	6 + 4			
1 + 5	2 + 5	3 + 5	4 + 5	5 + 5				
1 + 6	2 + 6	3 + 6	4 + 6					
1 + 7	2 + 7	3 + 7						
1 + 8	2 + 8							
1 + 9								

Make flash cards for the facts you need to practice.

5 + 4	9
front	back

Learn

Put each addition fact under the correct answer.

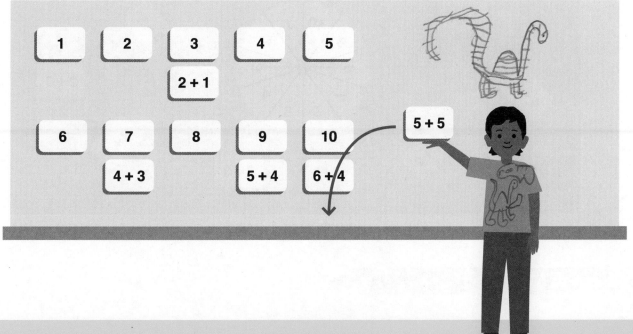

Do

1 What addition facts do you see in the ten-frame cards?

5 + 4 = 9

5 + 2 = 7

2 Find addition facts.

6 + 2 = 8

4 + 4 = 8

6 + 2

4 + 4

1 + 1

3 + 5

2 + 3

4 + 6

5 + 2

9 + 1

7 + 3

Exercise 7 • page 45

3-7 Addition Facts

1 Write an addition equation for each picture.

(a)

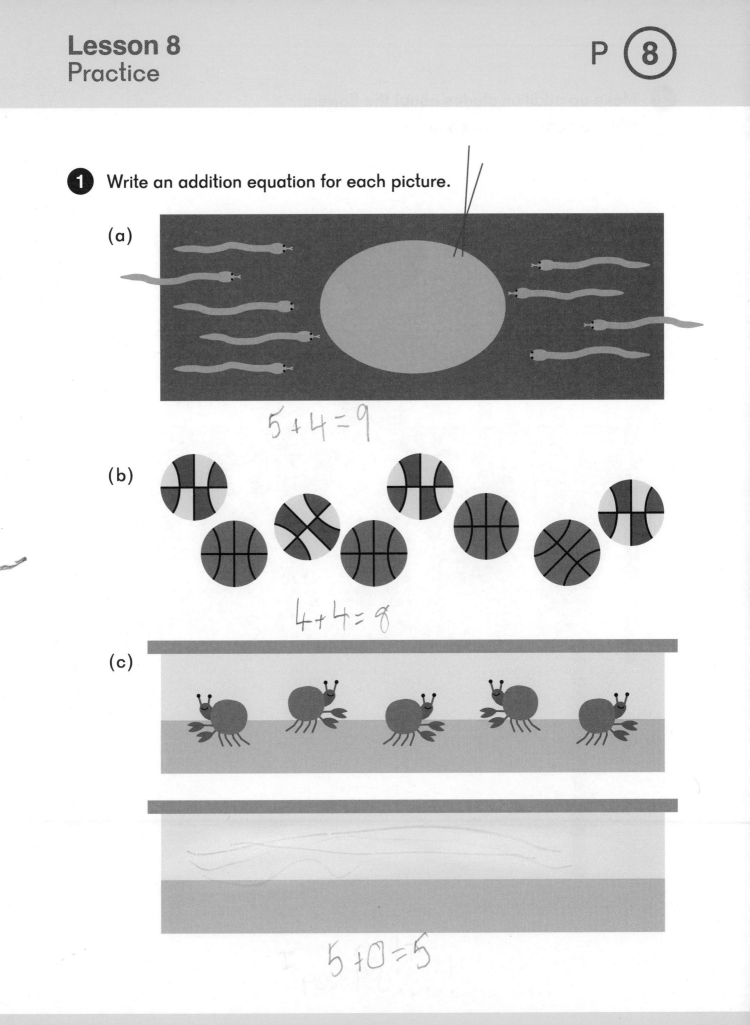

$$5 + 4 = 9$$

(b)

$$4 + 4 = 8$$

(c)

$$5 + 0 = 5$$

Make up addition stories about the flowers.
Write an equation for each.

2 + 3 = 6

2 + 3 = 5

1 + 2 = 3

3 (a)

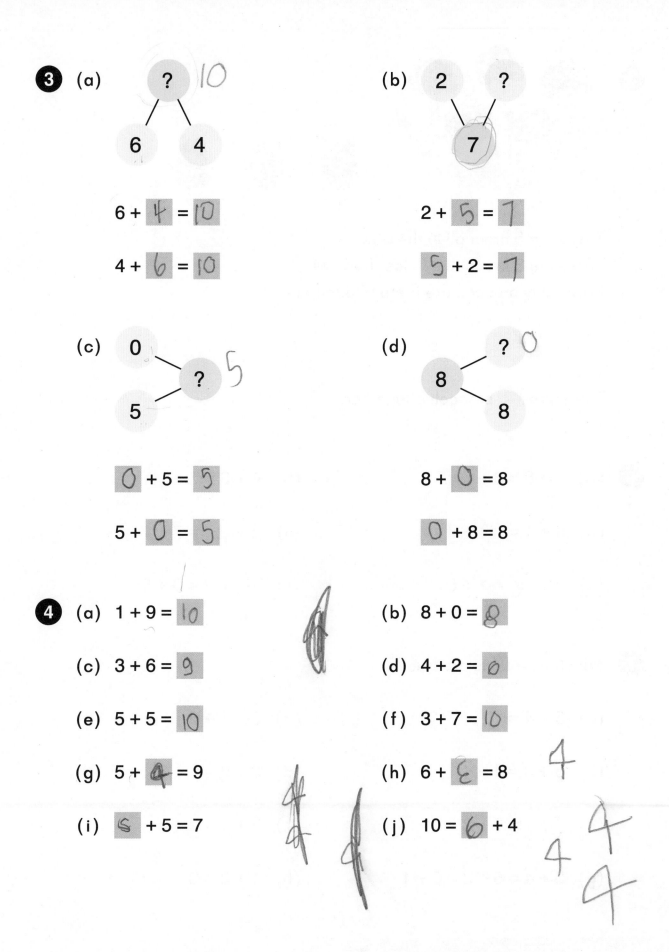

? 10

6   4

6 + 4 = 10

4 + 6 = 10

(b)

2   ?

7

2 + 5 = 7

5 + 2 = 7

(c)

0

? 5

5

0 + 5 = 5

5 + 0 = 5

(d)

? 0

8

8

8 + 0 = 8

0 + 8 = 8

4 (a) 1 + 9 = 10

(b) 8 + 0 = 8

(c) 3 + 6 = 9

(d) 4 + 2 = 6

(e) 5 + 5 = 10

(f) 3 + 7 = 10

(g) 5 + 4 = 9

(h) 6 + 2 = 8

(i) 2 + 5 = 7

(j) 10 = 6 + 4

5

There are 5 broccoli in the bowl.
There are 3 broccoli outside the bowl.
How many broccoli are there altogether?

$5 + \boxed{4} = \boxed{8}$

There are $\boxed{8}$ broccoli altogether.

6 (a) $2 + 8 = 8 + \boxed{2}$

(b) $4 + 5 = 7 + \boxed{2}$ ⁹

(c) $6 + 1 = \boxed{0} + 7$ ⁷

(d) $5 + \boxed{3} = 7 + 1$

(e) $3 + \boxed{X} = 1 + 6$

(f) $\boxed{3} + 2 = 0 + 5$

7 Which equations are true?

(a) $3 + 4 = 10$

(b) $9 = 3 + 6$

(c) $3 + 3 = 4 + 4$
    ₆

(d) $3 + 6 = 6 + 3$

(e) $4 + 1 = 5 + 2$

(f) $3 + 5 = 2 + 6$

(g) $3 + 4 = 5 + 2 = 6 + 1$
    ₇   ₇   ₇

(h) $1 + 2 = 3 + 4 = 7$

Exercise 8 • page 49

3-8 Practice

Chapter 4

Subtraction

There are 6 children.
2 children are leaving.
4 children are still in the park.

There are 9 balloons.
6 balloons are red.
3 balloons are yellow.

$6+3=9$

$6-2=4$

$5-2=3$

$3+2=5$

$2+7=9$

$9-2=7$

Think

There were 8 ants on a log.   8 − 5 = 3
5 ants left the log.
How many ants are still on the log? 3

Learn

 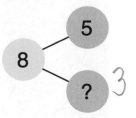

8 − 5 = 3

There are 3 ants still on the log.

This is **subtraction**.
— is read as minus.

Do

1

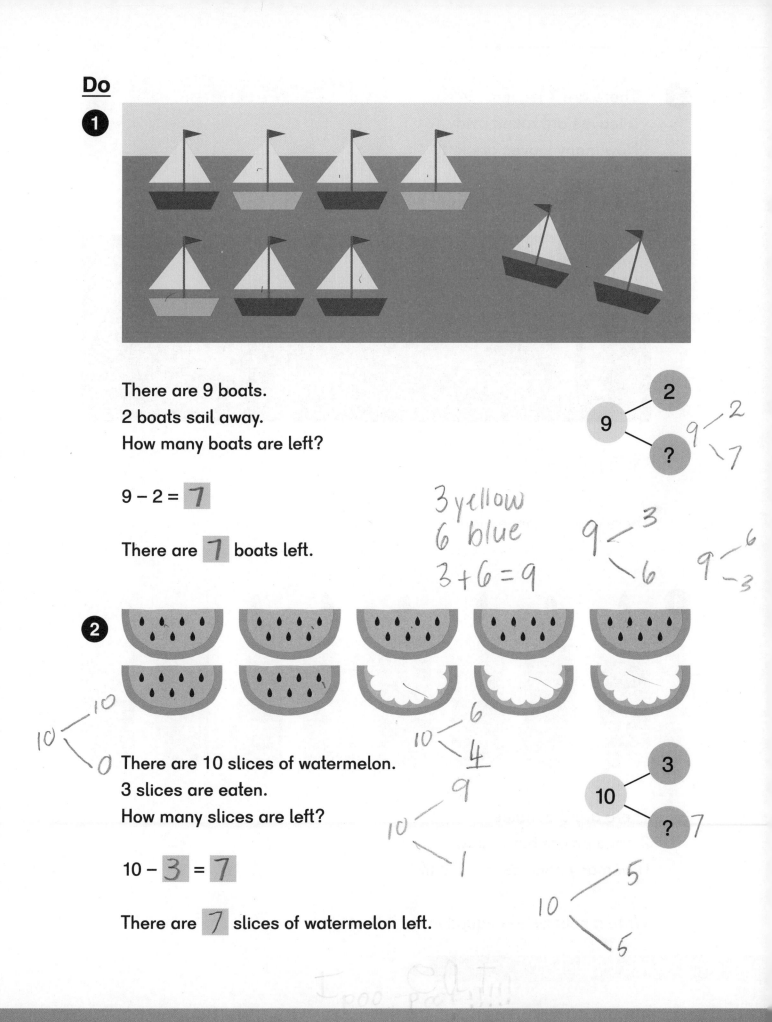

There are 9 boats.
2 boats sail away.
How many boats are left?

9 − 2 = 7

There are 7 boats left.

Handwritten:
3 yellow
6 blue
3 + 6 = 9

9 ⟨ 2 / 7
9 ⟨ 3 / 6
9 ⟨ 6 / 3

9 ⟨ 2 / ?

2

There are 10 slices of watermelon.
3 slices are eaten.
How many slices are left?

10 − 3 = 7

There are 7 slices of watermelon left.

Handwritten:
10 ⟨ 10 / 0
10 ⟨ 6 / 4
10 ⟨ 9 / 1
10 ⟨ 3 / ?
10 ⟨ 5 / 5

3 There are 7 leaves.
4 leaves are raked away.
How many leaves are left?

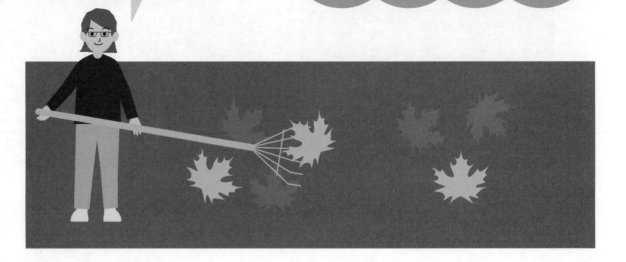

$7 - 4 = 3$

There are 3 leaves left.

4

$8 - 2 = 6$

There are 8 candles.
2 candles are blown out.
How many candles are still lit?

Write a subtraction equation for this story.

Exercise 1 • page 53

4-1 Subtraction as Taking Away

Think

There are 7 buckets.
3 buckets are blue.
How many buckets are pink?

Learn

7 – 3 = 4

There are 4 pink buckets.

Do

1

There are 6 owls.
1 owl is sleeping.
How many owls are awake?

$6 - 1 = 5$

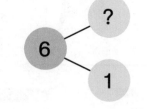

5 owls are awake.

2

There are 10 rulers.
7 rulers are red.
The rest are blue.
How many blue
rulers are there?

$10 - 7 = 3$

There are 3 blue rulers.

3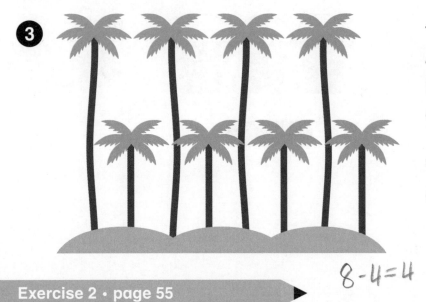

There are 8 palm trees.
4 palm trees are tall.
How many palm trees
are short?

Write a subtraction
equation for this story.

$8 - 4 = 4$

Exercise 2 • page 55

4-2 Subtraction as Taking Apart

Think

Sofia has 8 grapes.
How many grapes will be left if she eats 1 or 2 or 3 grapes?

Learn

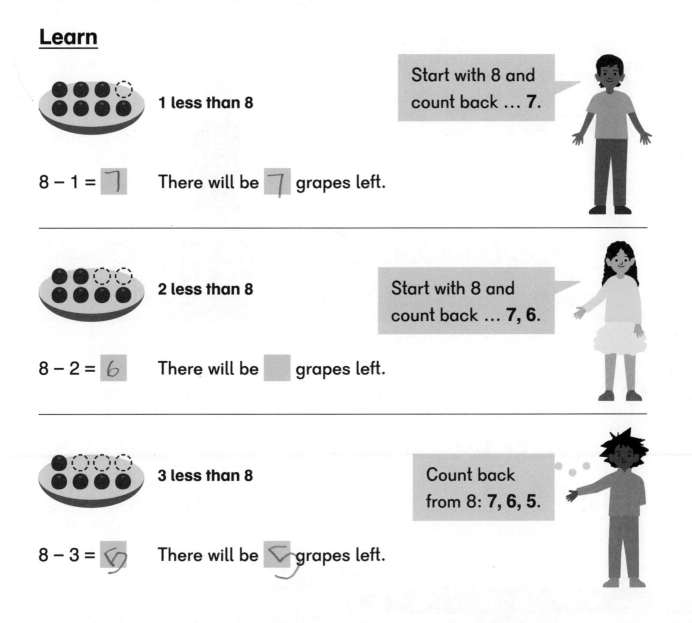

1 less than 8

Start with 8 and count back ... **7.**

8 − 1 = 7   There will be 7 grapes left.

2 less than 8

Start with 8 and count back ... **7, 6.**

8 − 2 = 6   There will be ☐ grapes left.

3 less than 8

Count back from 8: **7, 6, 5.**

8 − 3 = 5   There will be 5 grapes left.

Do

1 Subtract 2 from 7.

$7 - 2 =$ [5]

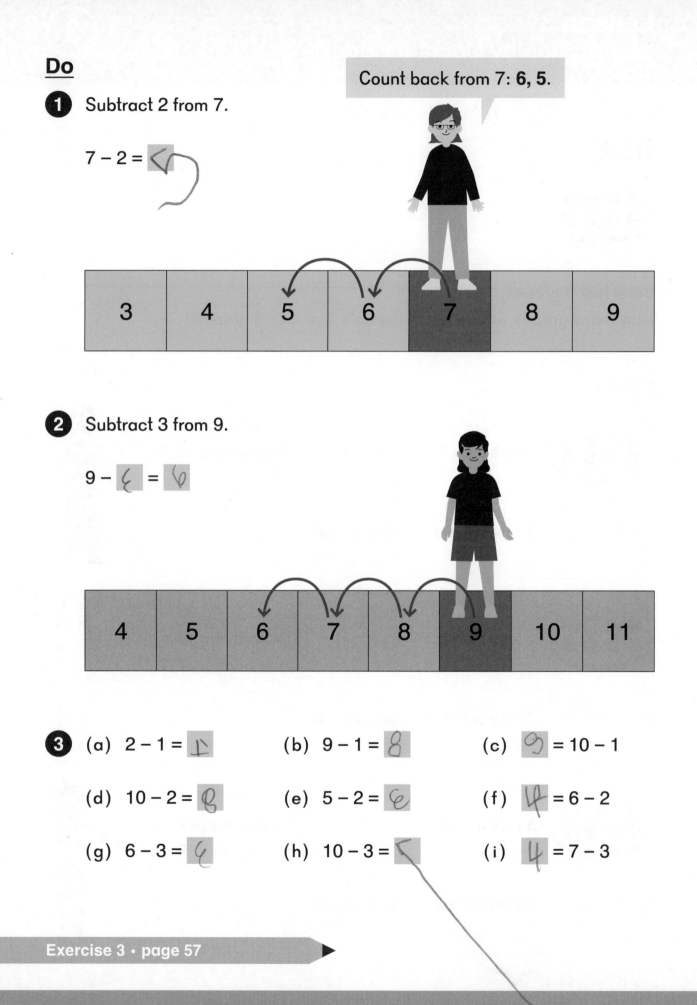

Count back from 7: **6, 5.**

| 3 | 4 | 5 | 6 | 7 | 8 | 9 |

2 Subtract 3 from 9.

$9 -$ [3] $=$ [6]

| 4 | 5 | 6 | 7 | 8 | 9 | 10 | 11 |

3 (a) $2 - 1 =$ [1]　　(b) $9 - 1 =$ [8]　　(c) [9] $= 10 - 1$

(d) $10 - 2 =$ [8]　　(e) $5 - 2 =$ [3]　　(f) [4] $= 6 - 2$

(g) $6 - 3 =$ [3]　　(h) $10 - 3 =$ [7]　　(i) [4] $= 7 - 3$

Exercise 3 • page 57

Think

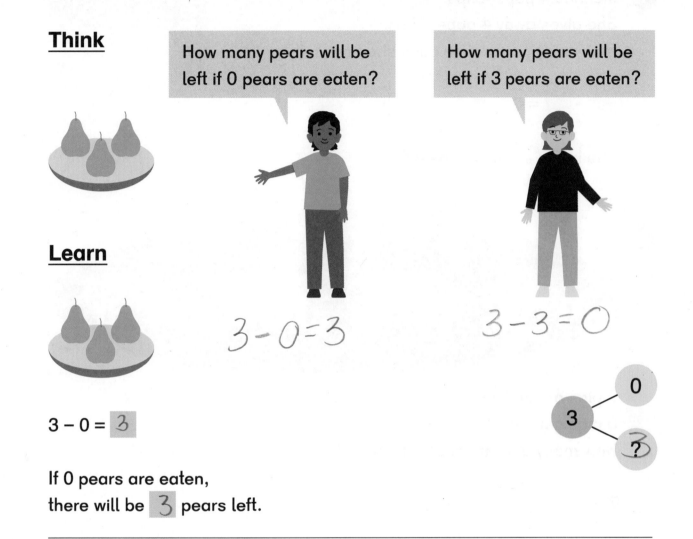

How many pears will be left if 0 pears are eaten?

How many pears will be left if 3 pears are eaten?

$3-0=3$

$3-3=0$

Learn

$3 - 0 = \boxed{3}$

If 0 pears are eaten,
there will be $\boxed{3}$ pears left.

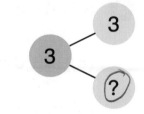

$3 - 3 = \boxed{0}$

If 3 pears are eaten,
there will be $\boxed{0}$ pears left.

Do

1 Mei has 4 paper clips.
She gives away 4 paper clips.
How many paper clips does she have left?

$4 - 4 = 0$

She has 0 paper clips left.

2

7 pigs are in the mud.
0 pigs get out.
How many pigs are still in the mud?

$7 - 0 = 7$

There are 7 pigs still in the mud.

3 (a) $6 - 6 = 0$ (b) $6 - 0 = 6$

(c) $8 - 0 = 8$ (d) $8 - 0 = 8$

(e) $10 - 0 = 10$ (f) $10 - 10 = 0$

(g) $0 - 0 = 0$ (h) $0 - 0 = 0$

Exercise 4 • page 59

4-4 Subtraction with 0

Think

Make up a subtraction story for the pens.

There are six markers. Two are being picked up. Four markers are left.

Learn

There are 6 pens.
2 pens are picked up.
There are 4 pens left in the tray.

6 − 2 = 4

6
2
? 4

There are 6 pens.

5 pens have caps.

1 pen does not have a cap.

6 − 5 = 1

?
?
?

What other subtraction stories can you make?

There are six markers. Three are green. three are red.

Do

1 Make up subtraction stories for each number sentence.

(a)

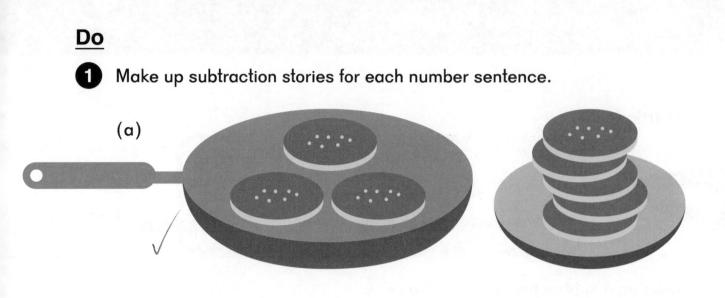

8 − 3 = 5

(b)

7 − 4 = 3

(c)

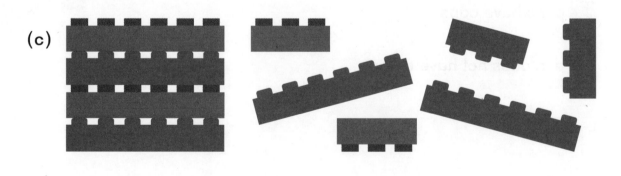

10 − 4 = 6

2 Make up different subtraction stories.
Write an equation for each story.

(a)

There are 8 slices of pizza.
1 slice has green peppers.
7 slices do not.

8 − 1 = 7

8−1=7

9−6=3

5=green
4=red

9−5=4

(b)

There are 9 dragonflies. one is on the flower.
8 are not on the flower.

9−1=8

(c)

3 have stars. 3 have stripes.
1 has dots.

There are 7 balloons.
3 are flying. 4 are not.
7−3=4

Think

There are 8 goats.
5 goats are brown.
How many goats are black?

There are 8 goats.
3 goats are black.
How many goats are brown?

Learn

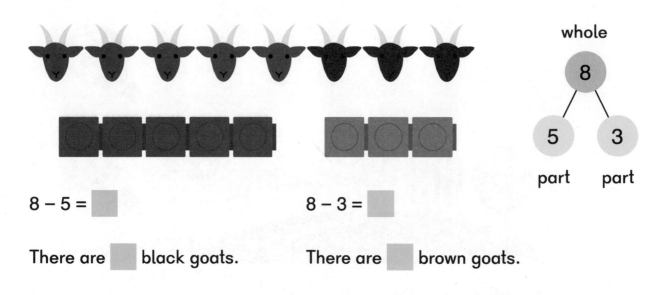

whole

8

5   3

part   part

8 − 5 = ▨

8 − 3 = ▨

There are ▨ black goats.

There are ▨ brown goats.

Do

1

Why are the 2 equations different?

There are 9 cupcakes.
5 cupcakes have sprinkles.
How many cupcakes
do not have sprinkles?

$9 - 5 = 4$

There are 9 cupcakes.
4 cupcakes do not have sprinkles.
How many cupcakes
have sprinkles?

$9 - 4 = 5$

2

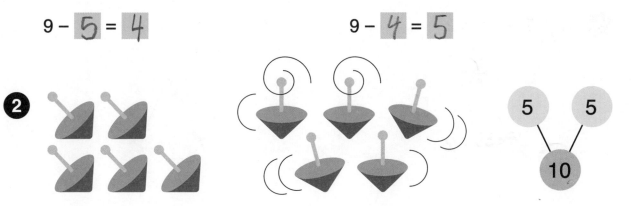

There are 10 tops.
5 tops are spinning.
How many tops are not spinning?

$10 - 5 = 5$

There are 10 tops.
5 tops are not spinning.
How many tops are spinning?

$10 - 5 = 5$

3 Make up 2 subtraction stories for the number bond.

Yvette
↓ maeve
Janna
Jacob
Cal
Harp

$6 - 2 = 4$
$6 - 4 = 2$

4 (a)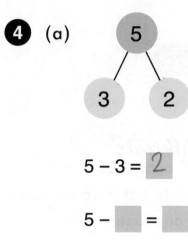

$5 - 3 = \boxed{2}$

$5 - \boxed{} = \boxed{}$

(b)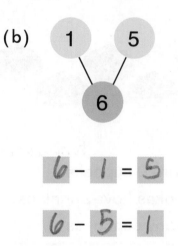

$\boxed{6} - \boxed{1} = \boxed{5}$

$\boxed{6} - \boxed{5} = \boxed{1}$

5 (a) $7 - \boxed{5} = 2$

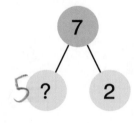

(b) $\boxed{9} - 7 = 2$

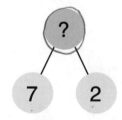

(c) $\boxed{} - 3 = 6$

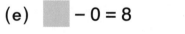
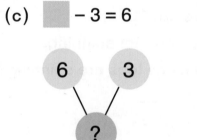

(d) $6 - \boxed{} = 3$

(e) $\boxed{} - 0 = 8$

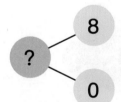

(f) $5 = 5 - \boxed{}$

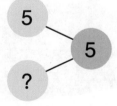

Exercise 6 • page 63

4-6 Subtraction with Number Bonds

Think

There are 7 sharpeners.
4 sharpeners are red.
3 sharpeners are purple.

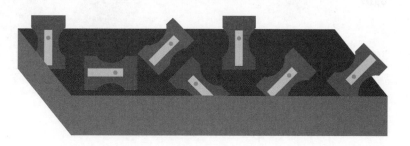

(a)   Write an equation for finding the total of 7 sharpeners. $3 + 4 = 7$

(b)   Write an equation for finding the 4 red sharpeners.

Learn

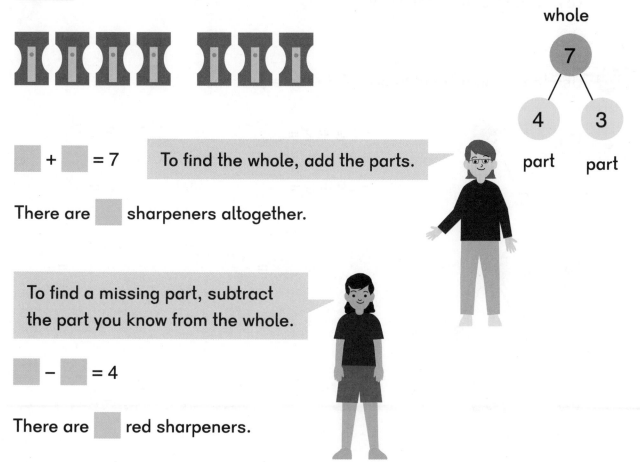

whole

7

4   3

part   part

☐ + ☐ = 7   To find the whole, add the parts.

There are ☐ sharpeners altogether.

To find a missing part, subtract
the part you know from the whole.

☐ − ☐ = 4

There are ☐ red sharpeners.

Write an equation for finding the 3 purple sharpeners.

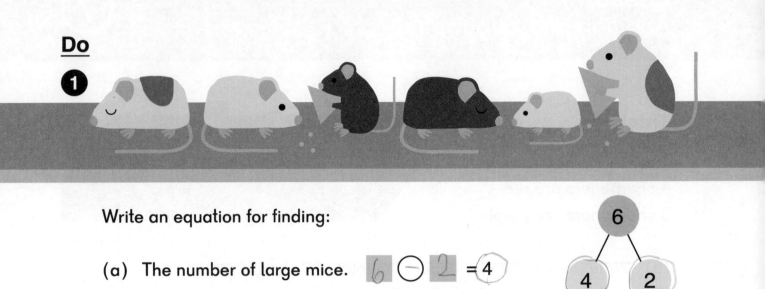

Write an equation for finding:

6

(a) The number of large mice. $6 \ominus 2 = 4$

(b) The number of small mice. $6 \ominus 4 = 2$

(c) The total number of mice. $6 \ominus 0 = 6$

4   2

2 Use the number bond to complete the equations.

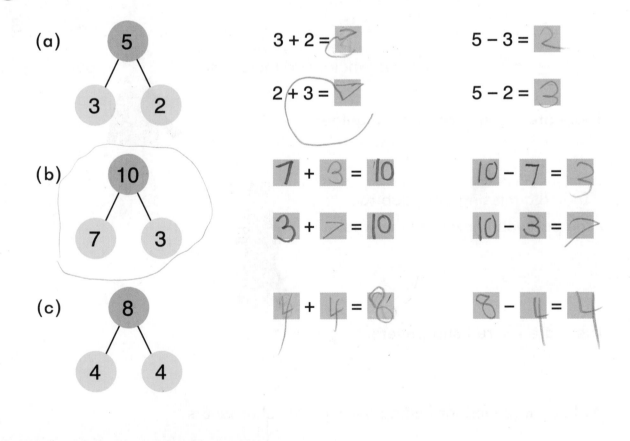

(a)

5

3   2

$3 + 2 = 5$

$2 + 3 = 5$

$5 - 3 = 2$

$5 - 2 = 3$

(b)

10

7   3

$7 + 3 = 10$

$3 + 7 = 10$

$10 - 7 = 3$

$10 - 3 = 7$

(c)

8

4   4

$4 + 4 = 8$

$8 - 4 = 4$

Exercise 7 • page 65

Think

Make addition and subtraction
story problems.
Write the equations.

I want to find how
many orange
slices there
are altogether.

☐ + ☐ = 10

I want to find how
many orange
slices are small.

☐ − ☐ = 4

Learn

4 slices of orange are small.
6 slices are big.
How many slices are
there altogether?

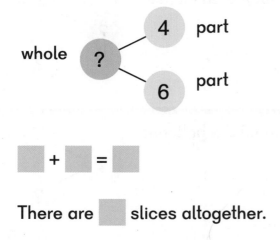

There are ☐ slices altogether.

There are 10 orange
slices altogether.
6 slices are big.
How many slices are small?

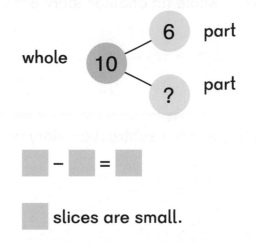

☐ slices are small.

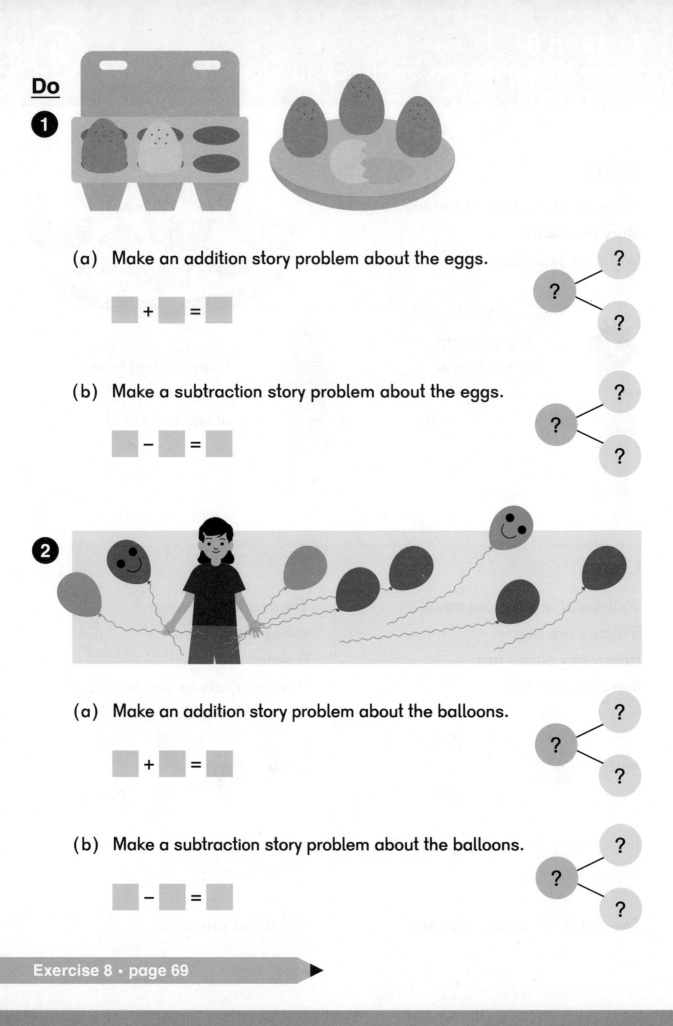

Do

1

(a) Make an addition story problem about the eggs.

⬜ + ⬜ = ⬜

(b) Make a subtraction story problem about the eggs.

⬜ – ⬜ = ⬜

2

(a) Make an addition story problem about the balloons.

⬜ + ⬜ = ⬜

(b) Make a subtraction story problem about the balloons.

⬜ – ⬜ = ⬜

Exercise 8 • page 69

Think

Say the answers and look for patterns.

2 – 1	3 – 1	4 – 1	5 – 1	6 – 1	7 – 1	8 – 1	9 – 1	10 – 1
	3 – 2	4 – 2	5 – 2	6 – 2	7 – 2	8 – 2	9 – 2	10 – 2
		4 – 3	5 – 3	6 – 3	7 – 3	8 – 3	9 – 3	10 – 3
			5 – 4	6 – 4	7 – 4	8 – 4	9 – 4	10 – 4
				6 – 5	7 – 5	8 – 5	9 – 5	10 – 5
					7 – 6	8 – 6	9 – 6	10 – 6
						8 – 7	9 – 7	10 – 7
							9 – 8	10 – 8
								10 – 9

Make flash cards for the facts you need to practice.

| 7 – 3 | 4 |
| front | back |

Learn

Put each subtraction fact under the correct answer.

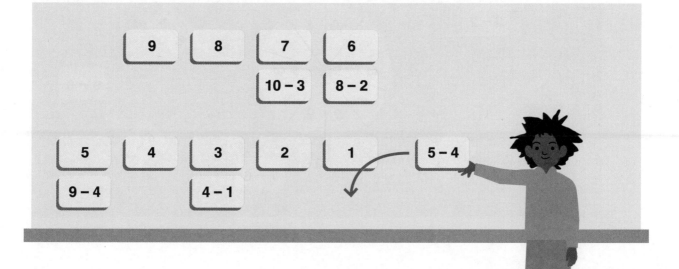

Do

1 What subtraction facts do you see in the ten-frame cards?

2 Find subtraction facts.

Exercise 9 · page 73

4-9 Subtraction Facts

1 Write a subtraction equation for each picture.

(a)

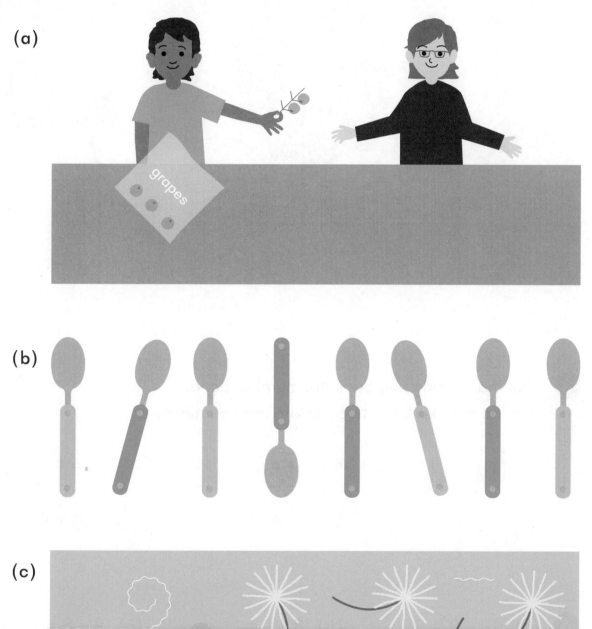

(b)

(c)

2 Make up subtraction stories about the bats.
Write an equation for each.

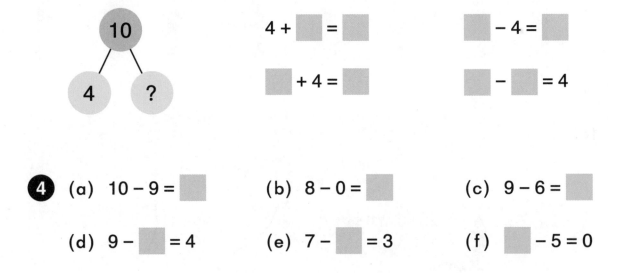

3 Find the missing number in the number bond.
Use the number bond to complete the equations.

10

4   ?

4 + ☐ = ☐   ☐ − 4 = ☐

☐ + 4 = ☐   ☐ − ☐ = 4

4 (a) 10 − 9 = ☐   (b) 8 − 0 = ☐   (c) 9 − 6 = ☐

(d) 9 − ☐ = 4   (e) 7 − ☐ = 3   (f) ☐ − 5 = 0

5 (a) $8 - 2 = 9 - \boxed{}$ (b) $9 - 5 = 8 - \boxed{}$

(c) $6 - 4 = \boxed{} - 6$ (d) $8 - \boxed{} = 5 - 1$

(e) $3 - \boxed{} = 6 - 6$ (f) $\boxed{} - 0 = 9 - 5$

6

There are 8 pineapples in all.
There are 3 pineapples outside the bag.
How many pineapples are in the bag?

$\boxed{} - \boxed{} = \boxed{}$

There are pineapples in the bag.

7 Which equations are false?

(a) $10 - 4 = 6$ (b) $6 - 2 = 8$

(c) $6 = 9 - 3$ (d) $3 - 3 = 4 - 4$

(e) $6 - 0 = 6 - 6$ (f) $7 - 2 = 9 - 2$

Exercise 10 • page 77

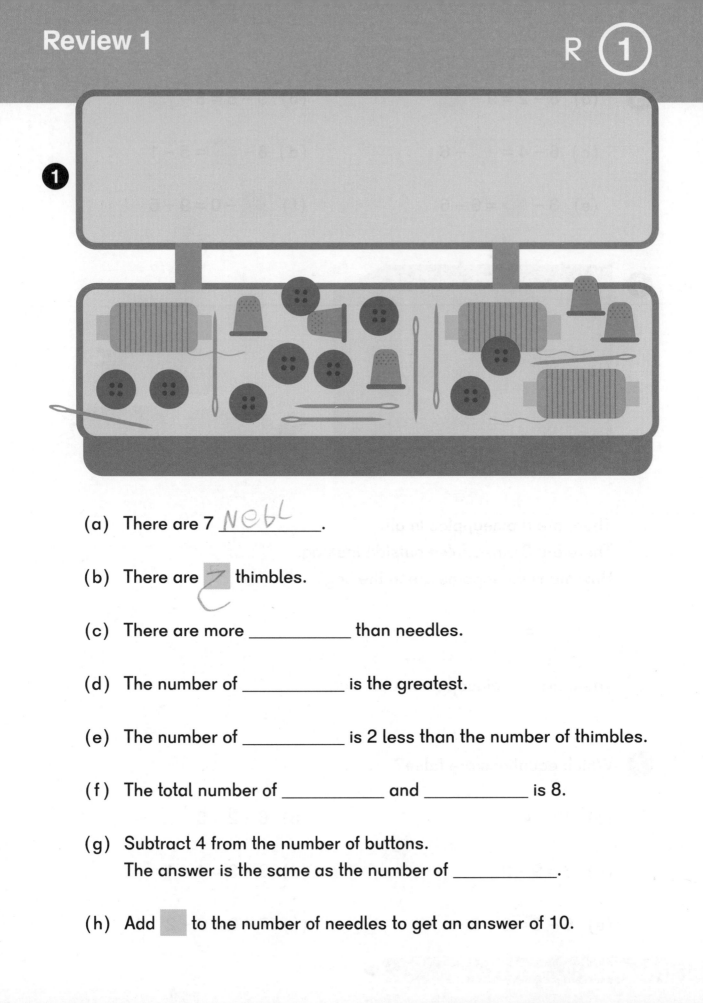

1

(a) There are 7 _Nebl_.

(b) There are 7 thimbles.

(c) There are more _____ than needles.

(d) The number of _____ is the greatest.

(e) The number of _____ is 2 less than the number of thimbles.

(f) The total number of _____ and _____ is 8.

(g) Subtract 4 from the number of buttons.
     The answer is the same as the number of _____.

(h) Add ▢ to the number of needles to get an answer of 10.

2 Complete the number bonds.

(a)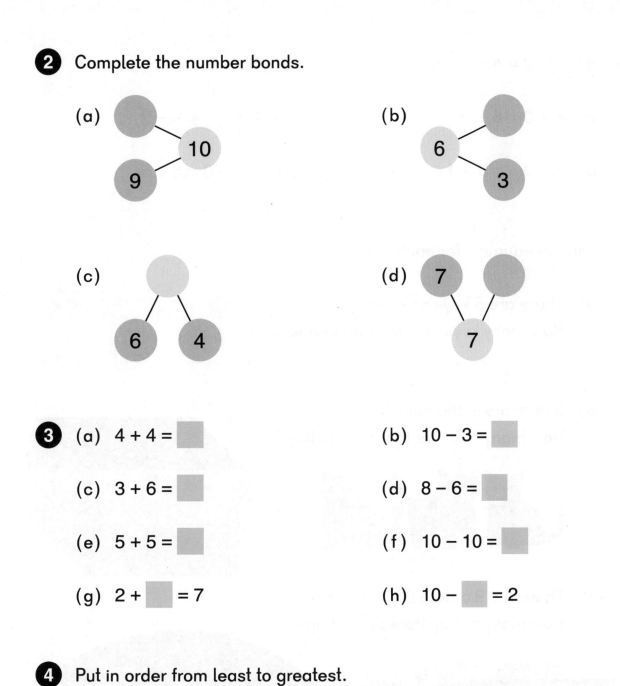

9 10

(b)

6 3

(c)

6 4

(d)

7 7 7

3 (a) $4 + 4 = \boxed{}$

(b) $10 - 3 = \boxed{}$

(c) $3 + 6 = \boxed{}$

(d) $8 - 6 = \boxed{}$

(e) $5 + 5 = \boxed{}$

(f) $10 - 10 = \boxed{}$

(g) $2 + \boxed{} = 7$

(h) $10 - \boxed{} = 2$

4 Put in order from least to greatest.

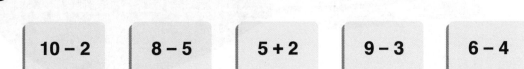

| $10 - 2$ | $8 - 5$ | $5 + 2$ | $9 - 3$ | $6 - 4$ |

5 (a) $8 - 2 = 4 + \square$

(b) $10 - 3 = 3 + \square$

(c) $6 - 0 = 6 + \square$

(d) $2 + \square = 9 - 2$

(e) $1 + \square = 8 - 1$

(f) $\square + 0 = 5 - 5$

6 Write an equation for each.

(a) There are 5 keys on a ring.
How many keys are there altogether?

(b) 3 cats are in the box.
How many cats are there altogether?

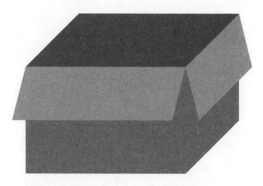

(c) There are 9 caterpillars altogether.
How many caterpillars are hiding?

(d) The necklace will have 10 beads.
How many more beads are needed?

Chapter 5

Numbers to 20

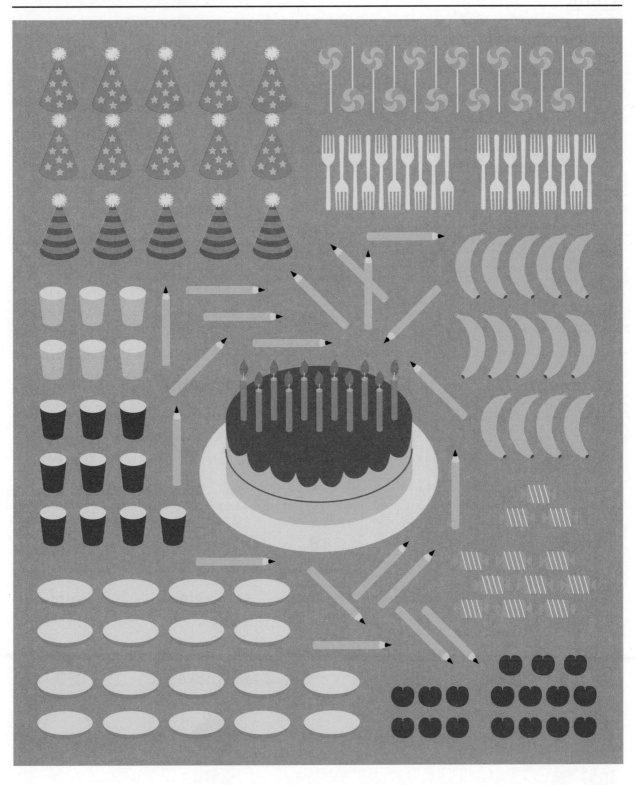

Think

How many pencils are there?

Learn

11	12	13	14	15
eleven	twelve	thirteen	fourteen	fifteen

16	17	18	19	20
sixteen	seventeen	eighteen	nineteen	twenty

17 is 10 and 7.

17
10 7

2 tens make 20.

20
10 10

<u>Do</u>

 Show the numbers from 11 to 20 with ten-frame cards, number cards, and number word cards.

| 11 | 12 | ... | 20 |

eleven   twelve   twenty

2 How many party hats are there?

10 and 5 make [ ].

10 + 5 = [ ]

3 How many lollipops are there?

10 and 3 make [ ].

10 + 3 = [ ]

4 How many cherries are there?

[ ] is 10 and 7.

[ ] = 10 + 7

Exercise 1 · page 83

Think

There are 10 pink cups and 6 purple cups.
How many cups are there altogether?

There are 16 cups.
6 cups are purple.
The rest are pink.
How many pink
cups are there?

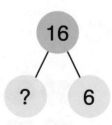

Learn

10 + 6 = ▢

There are ▢ cups altogether.

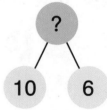

16 − 6 = ▢

There are ▢ pink cups.

1

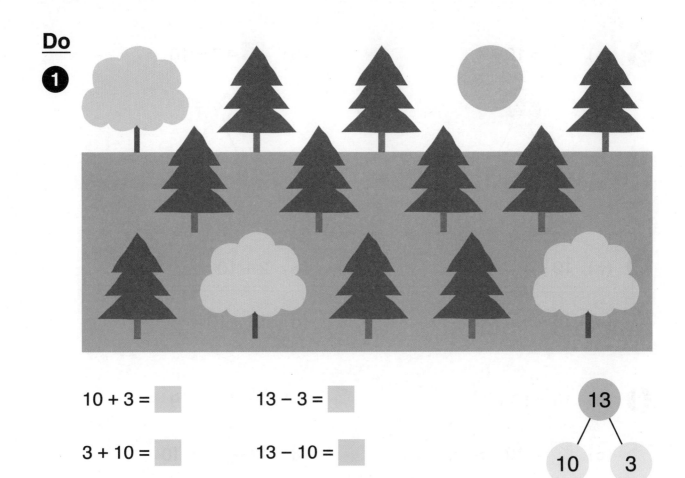

10 + 3 = ▢   13 − 3 = ▢

3 + 10 = ▢   13 − 10 = ▢

13
10   3

2

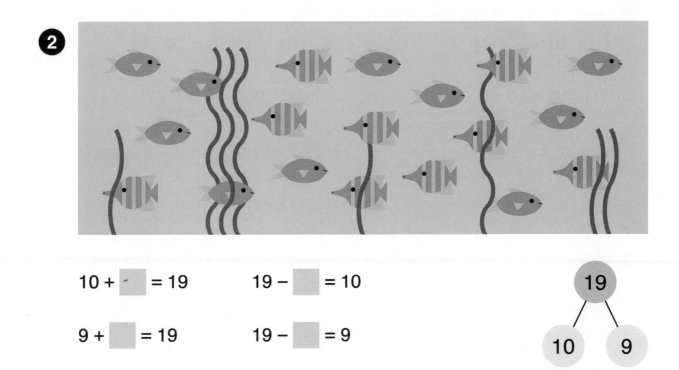

10 + ▢ = 19   19 − ▢ = 10

9 + ▢ = 19   19 − ▢ = 9

19
10   9

3 (a) $\boxed{} = 10 + 4$

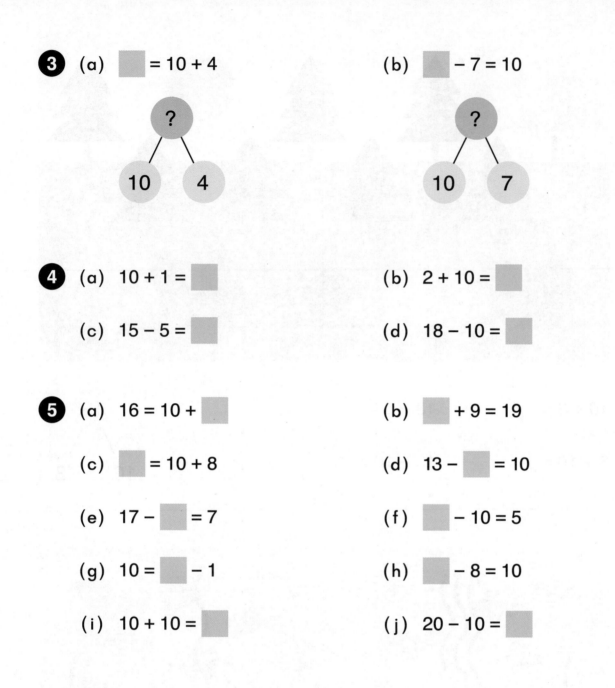

(b) $\boxed{} - 7 = 10$

4 (a) $10 + 1 = \boxed{}$

(b) $2 + 10 = \boxed{}$

(c) $15 - 5 = \boxed{}$

(d) $18 - 10 = \boxed{}$

5 (a) $16 = 10 + \boxed{}$

(b) $\boxed{} + 9 = 19$

(c) $\boxed{} = 10 + 8$

(d) $13 - \boxed{} = 10$

(e) $17 - \boxed{} = 7$

(f) $\boxed{} - 10 = 5$

(g) $10 = \boxed{} - 1$

(h) $\boxed{} - 8 = 10$

(i) $10 + 10 = \boxed{}$

(j) $20 - 10 = \boxed{}$

Exercise 2 · page 85

5-2 Add or Subtract Tens or Ones

Think

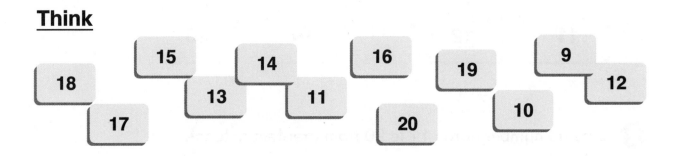

How can we order these numbers?

Learn

Count up from 0 to 20.
Count down from 20 to 0.

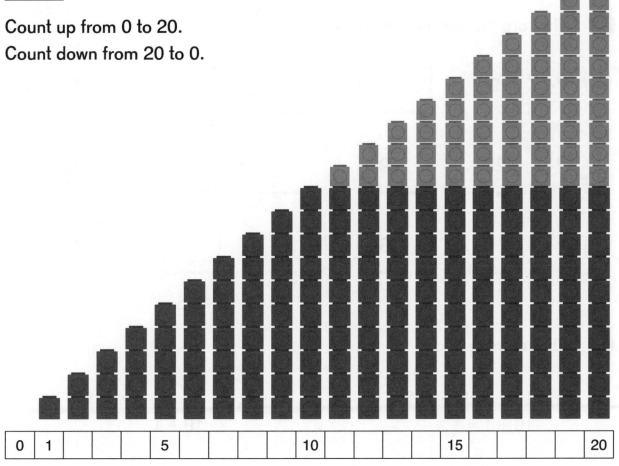

| 0 | 1 | | | | 5 | | | | | 10 | | | | | 15 | | | | | 20 |

What are the missing numbers?

Do

1 Line up number cards 11 to 20 from least to greatest.

| 11 | 12 | ... | 19 | 20 |

2 Line up number cards 11 to 20 from greatest to least.

| 20 | 19 | ... | 12 | 11 |

3 What are the missing numbers?

(a) | 9 | | 11 | | 13 |

(b) | 20 | 19 | | | 16 |

(c) | | 15 | | | 18 |

4 What number is...

(a) 1 more than 10?
$10 + 1 = \boxed{}$

(b) 1 more than 19?
$19 + 1 = \boxed{}$

(c) 1 less than 15?
$15 - 1 = \boxed{}$

(d) 1 less than 20?
$20 - 1 = \boxed{}$

5 What number is 2 more than 18?

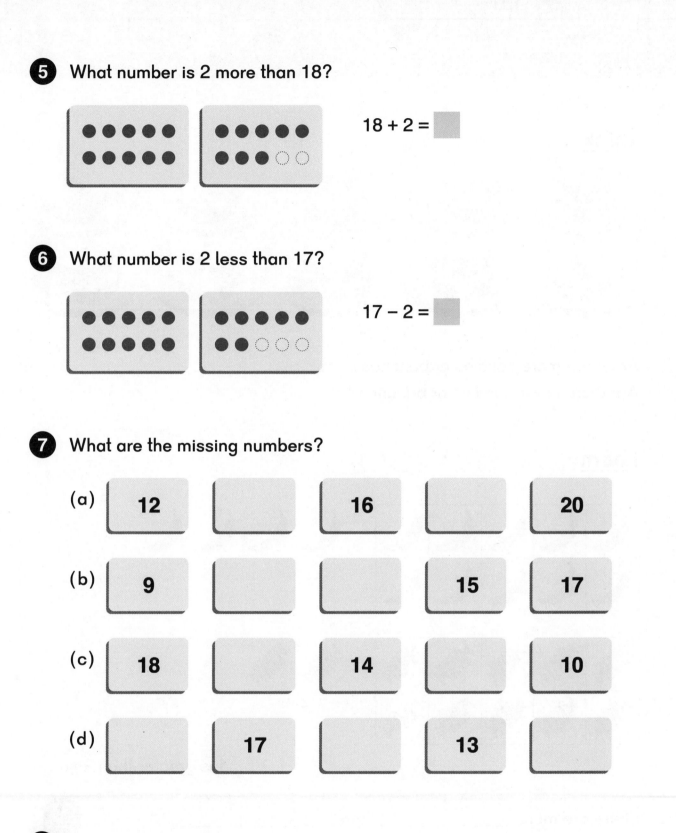

$18 + 2 = \boxed{}$

6 What number is 2 less than 17?

$17 - 2 = \boxed{}$

7 What are the missing numbers?

(a) | 12 | | 16 | | 20 |

(b) | 9 | | | 15 | 17 |

(c) | 18 | | 14 | | 10 |

(d) | | 17 | | 13 | |

8 Play with a partner.
Show a number card and say the number that is 2 more and 2 less.

Exercise 3 • page 87

Think

Are there more candies or bananas?

Are there fewer candies or bananas?

Learn

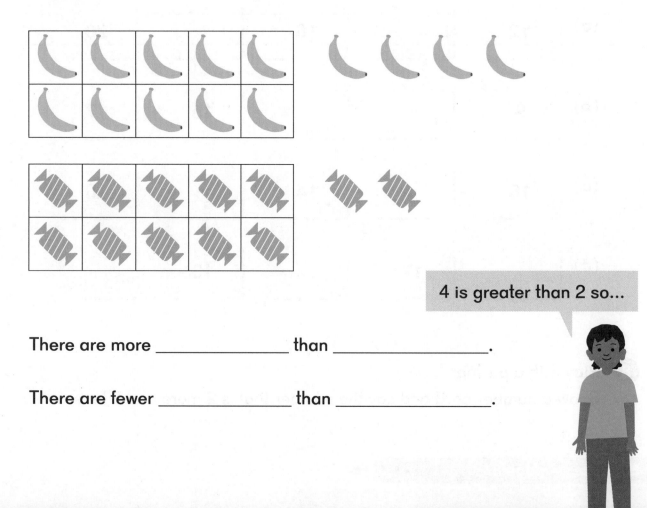

4 is greater than 2 so...

There are more _____ than _____.

There are fewer _____ than _____.

<u>Do</u>

1 Which is greater?

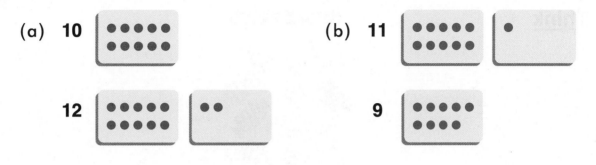

(a) 10

12

(b) 11

9

2 Which is less?

(a) 16

19

(b) 20

17

3 Compare these numbers.

| 14 | 8 | 19 | 15 |

(a) Which is the greatest?

(b) Which is the least?

Exercise 4 • page 89

Think

There are 12 green candies and 3 orange candies.
How many candies are there altogether?

Learn

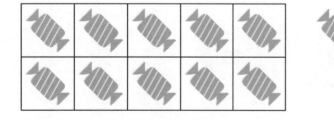

12 + 3 =

There are ☐ candies altogether.

2 + 3 = 5

Do

1 Add 15 and 3.

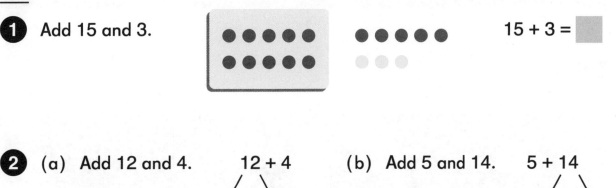

$15 + 3 = \boxed{}$

2 (a) Add 12 and 4.

$12 + 4 = \boxed{}$

$12 + 4$
10   2

(b) Add 5 and 14.

$5 + 14 = \boxed{}$

$5 + 14$
4   10

3

11 basketballs are in the bag.

4 more basketballs are added.

How many basketballs are in the bag now?

$\boxed{} + \boxed{} = \boxed{}$

There are $\boxed{}$ basketballs in the bag now.

4 (a) $16 + 3 = \boxed{}$

(b) $12 + 7 = \boxed{}$

(c) $5 + 11 = \boxed{}$

(d) $3 + 13 = \boxed{}$

(e) $\boxed{} + 10 = 20$

(f) $17 = 15 + \boxed{}$

Exercise 5 • page 91

Think

There are 15 foxes sitting in the snow.

3 foxes go away to find something to eat.

How many foxes are left?

Learn

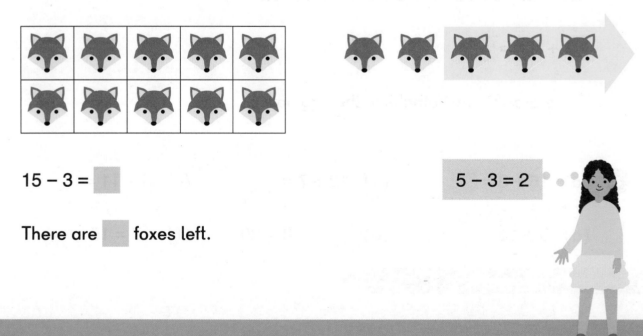

$15 - 3 =$

There are ___ foxes left.

$5 - 3 = 2$

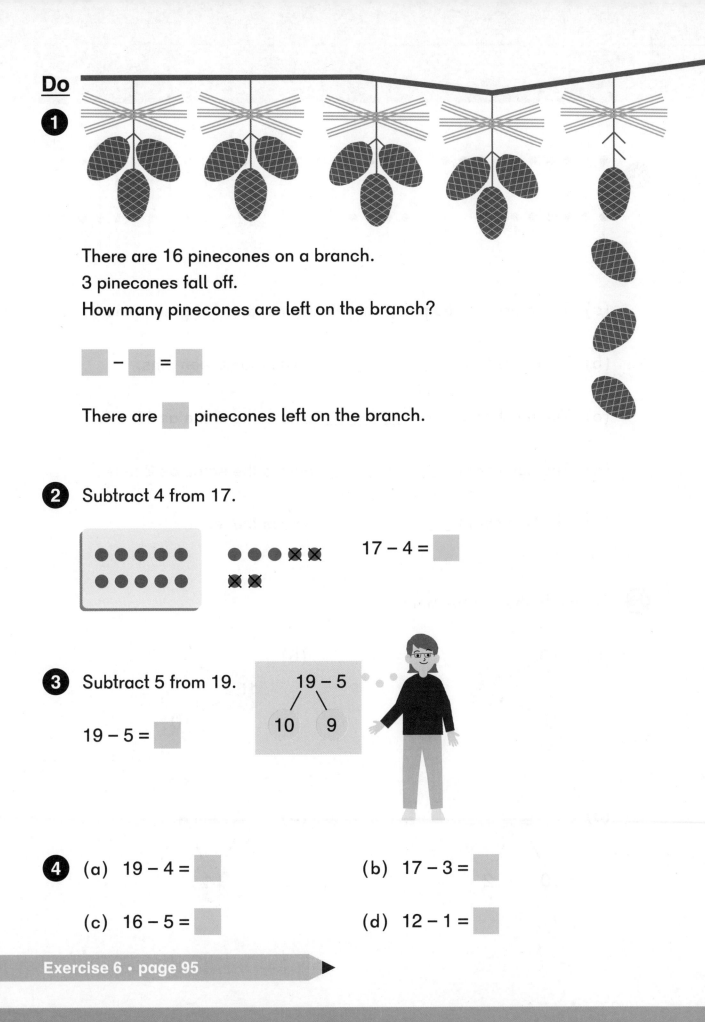

Do

1

There are 16 pinecones on a branch.
3 pinecones fall off.
How many pinecones are left on the branch?

☐ – ☐ = ☐

There are ☐ pinecones left on the branch.

2 Subtract 4 from 17.

$17 - 4 = $ ☐

3 Subtract 5 from 19.

$19 - 5 = $ ☐

$$19 - 5$$
$$10 \qquad 9$$

4 (a) $19 - 4 = $ ☐   (b) $17 - 3 = $ ☐

(c) $16 - 5 = $ ☐   (d) $12 - 1 = $ ☐

Exercise 6 • page 95

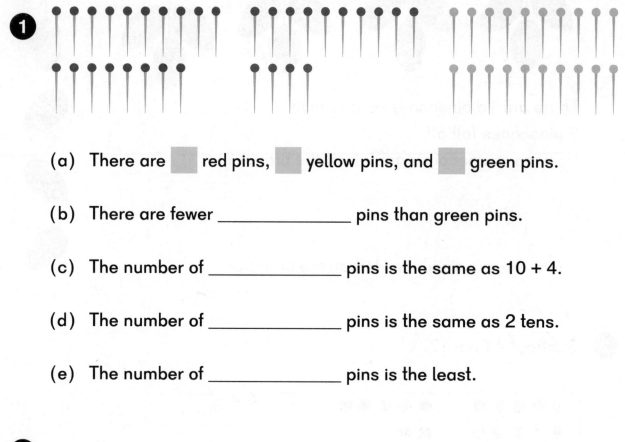

1

(a) There are ☐ red pins, ☐ yellow pins, and ☐ green pins.

(b) There are fewer _____ pins than green pins.

(c) The number of _____ pins is the same as 10 + 4.

(d) The number of _____ pins is the same as 2 tens.

(e) The number of _____ pins is the least.

2 Complete the number bonds.

(a) 10 — ◯
     9 — ╱

(b) ◯
     16 —
     10

(c) ◯
     10   2

(d) ◯   4
        14

3 Write the numbers.

(a) eleven

(b) thirteen

(c) eighteen

(d) twelve

(e) twenty

(f) fifteen

4 Put in order from least to greatest.

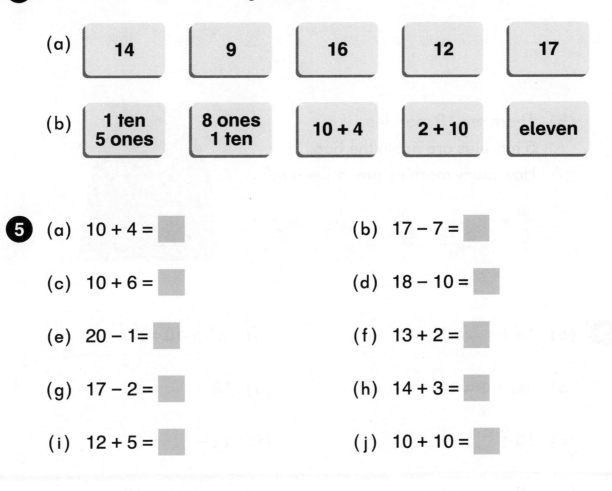

(a)

| 14 | 9 | 16 | 12 | 17 |

(b)

| 1 ten 5 ones | 8 ones 1 ten | 10 + 4 | 2 + 10 | eleven |

5 (a) 10 + 4 = ☐

(b) 17 − 7 = ☐

(c) 10 + 6 = ☐

(d) 18 − 10 = ☐

(e) 20 − 1 = ☐

(f) 13 + 2 = ☐

(g) 17 − 2 = ☐

(h) 14 + 3 = ☐

(i) 12 + 5 = ☐

(j) 10 + 10 = ☐

Write an equation for each and find the answer.

(a) There are 15 cookies in the box.
If 3 more cookies are put in the box,
how many cookies will be in the box?

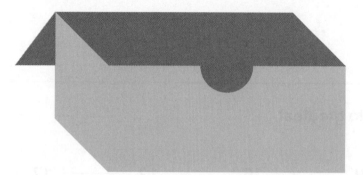

(b) There are 19 marbles.
5 marbles are not in the bag.
How many marbles are in the bag?

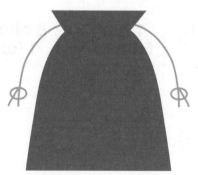

7 (a) 14 + ▨ = 16   (b) ▨ + 10 = 20

(c) ▨ + 1 = 11   (d) 19 − ▨ = 10

(e) 10 + ▨ = 15   (f) 13 − ▨ = 3

(g) 18 − ▨ = 12   (h) ▨ + 13 = 18

(i) 16 − ▨ = 11   (j) 12 + ▨ = 19

Exercise 7 • page 99

Chapter 6

Addition to 20

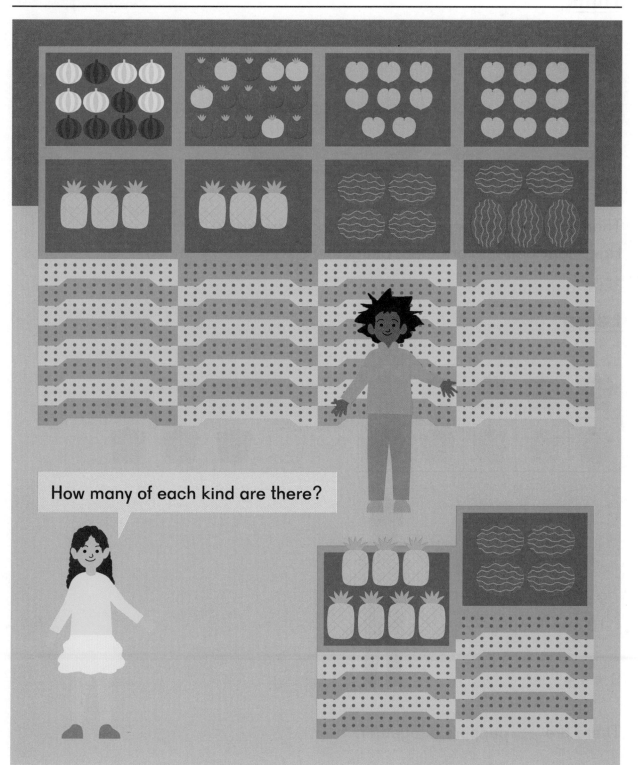

How many of each kind are there?

Think

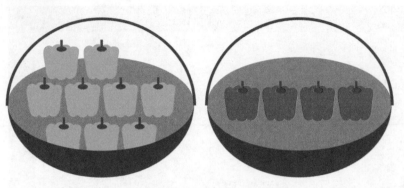

Make a 10!

There are 9 yellow peppers and 4 orange peppers.
How many peppers are there altogether?

Learn

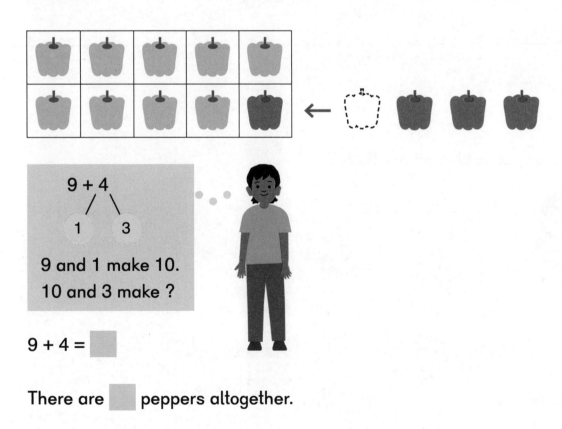

9 + 4
1   3

9 and 1 make 10.
10 and 3 make ?

9 + 4 = ▢

There are ▢ peppers altogether.

Do

1

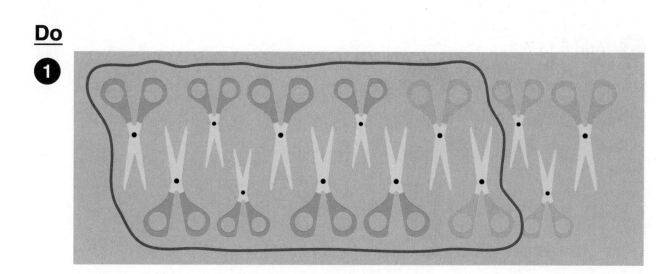

There are 8 green scissors and 5 pink scissors.
How many scissors are there in all?

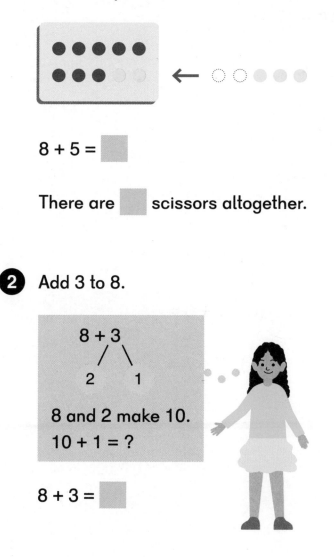

$8 + 5 = $

There are ▢ scissors altogether.

2 Add 3 to 8.

$8 + 3$

2   1

8 and 2 make 10.
$10 + 1 = ?$

$8 + 3 = $

3 Add 5 to 7.

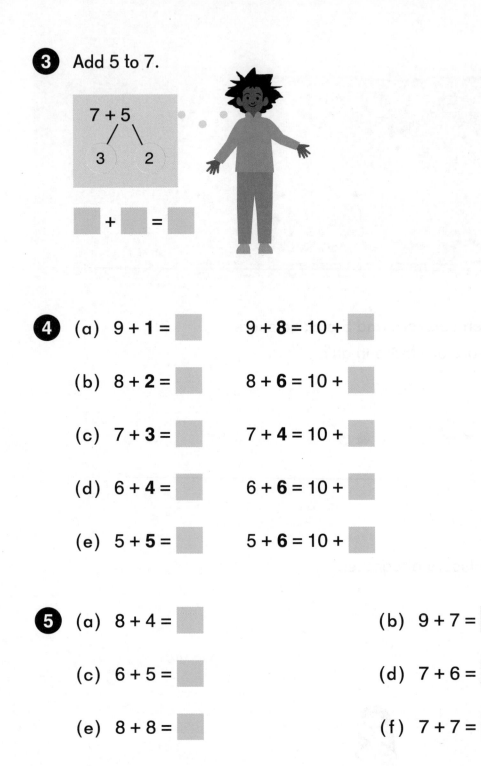

7 + 5

3   2

▢ + ▢ = ▢

4 (a) 9 + 1 = ▢   9 + **8** = 10 + ▢

   (b) 8 + **2** = ▢   8 + **6** = 10 + ▢

   (c) 7 + **3** = ▢   7 + **4** = 10 + ▢

   (d) 6 + **4** = ▢   6 + **6** = 10 + ▢

   (e) 5 + **5** = ▢   5 + **6** = 10 + ▢

5 (a) 8 + 4 = ▢   (b) 9 + 7 = ▢

   (c) 6 + 5 = ▢   (d) 7 + 6 = ▢

   (e) 8 + 8 = ▢   (f) 7 + 7 = ▢

Exercise 1 • page 101

Think

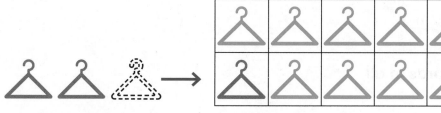

Sofia has 3 orange coat hangers
and 9 blue coat hangers.
How many coat hangers does
she have altogether?

Learn

$$3 + 9$$

2   1

1 and 9 make 10.
2 and 10 make ?

3 + 9 = ▢

She has ▢ coat hangers altogether.

Do

1

There are 4 green lizards and 7 blue lizards.
How many lizards are there in all?

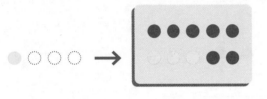

4 + 7 = ⬜

There are ⬜ lizards in all.

2 Add 8 to 4.

4 + 8
/ \
2   2

2 and 8 make 10.
2 + 10 = ?

4 + 8 = ⬜

3 Add 7 to 6.

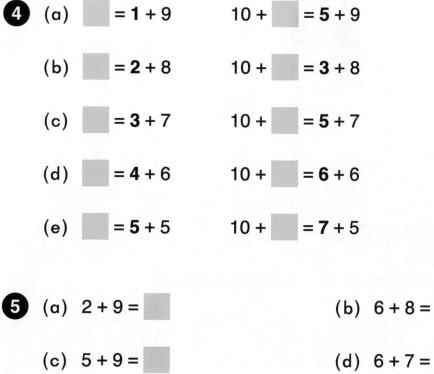

$6 + 7 = \boxed{}$

4 (a) $\boxed{} = 1 + 9$ $10 + \boxed{} = 5 + 9$

   (b) $\boxed{} = 2 + 8$ $10 + \boxed{} = 3 + 8$

   (c) $\boxed{} = 3 + 7$ $10 + \boxed{} = 5 + 7$

   (d) $\boxed{} = 4 + 6$ $10 + \boxed{} = 6 + 6$

   (e) $\boxed{} = 5 + 5$ $10 + \boxed{} = 7 + 5$

5 (a) $2 + 9 = \boxed{}$ (b) $6 + 8 = \boxed{}$

   (c) $5 + 9 = \boxed{}$ (d) $6 + 7 = \boxed{}$

   (e) $5 + 8 = \boxed{}$ (f) $4 + 7 = \boxed{}$

Exercise 2 • page 105

Think

One box has 7 chocolates.
The other box has 8 chocolates.

How many chocolates are there altogether?

Learn

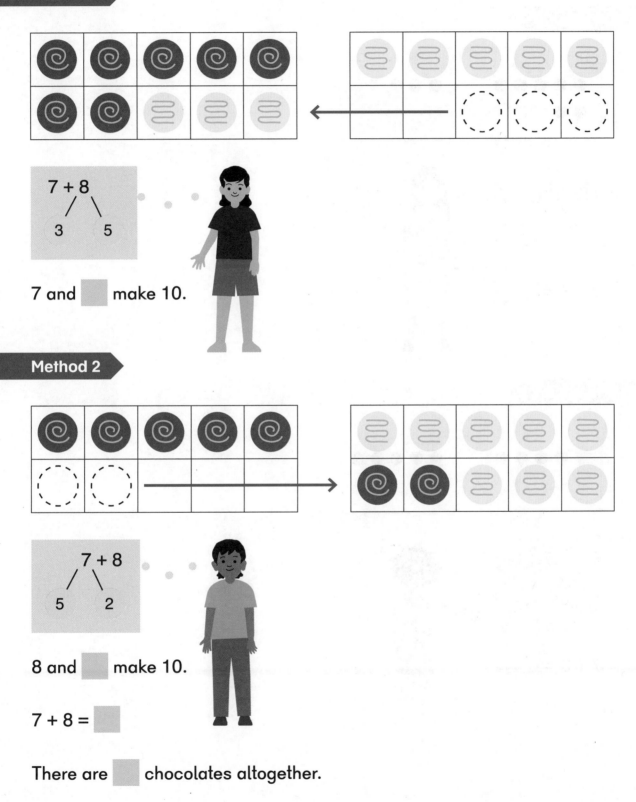

7 + 8

3   5

7 and ⬜ make 10.

Method 2

7 + 8

5   2

8 and ⬜ make 10.

7 + 8 = ⬜

There are ⬜ chocolates altogether.

Do

1 Add 7 and 6.

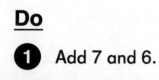

Method 1

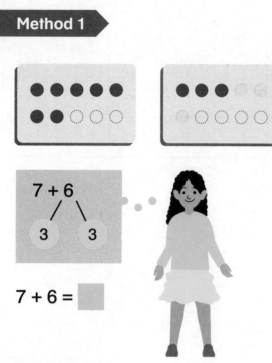

7 + 6

3   3

7 + 6 = ▢

Method 2

7 + 6

3   4

7 + 6 = ▢

2 Add 5 and 6.

Method 1

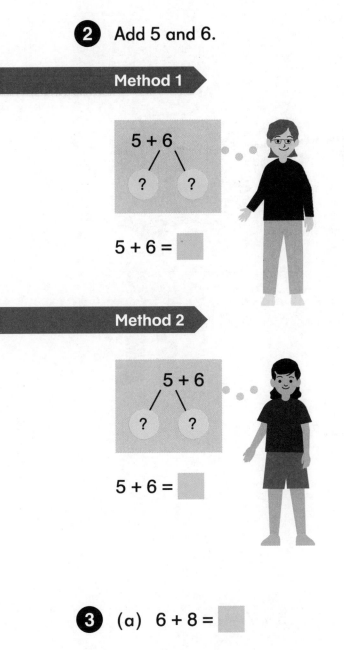

5 + 6

? ?

5 + 6 = ☐

Method 2

5 + 6

? ?

5 + 6 = ☐

3 (a) 6 + 8 = ☐   (b) 7 + 9 = ☐

(c) 6 + 6 = ☐   (d) 8 + 8 = ☐

(e) 9 + 9 = ☐   (f) 7 + 7 = ☐

Dion picks 7 blueberries.
Then he picks 6 more blueberries.
How many blueberries does he have now?

He has blueberries now.

 (a) 7 more than 5 is ☐.

(b) 6 more than 8 is ☐.

(c) 9 more than 5 is ☐.

$5 + 7 = ?$

Exercise 3 • page 109

Lesson 4
Addition Facts to 20

Think

Say the answers and look for patterns.

Make flash cards for the facts you need to practice.

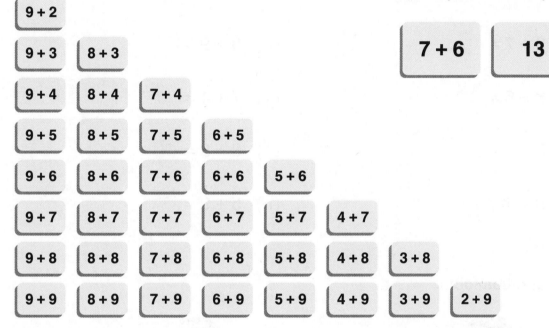

Learn

Put each addition fact under the correct answer.

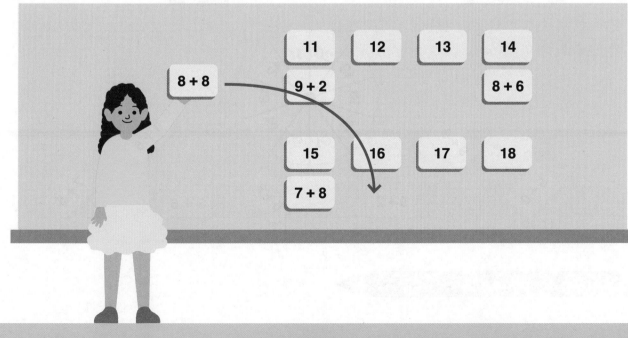

Do

1 (a) $8 + 5 = \boxed{}$ (b) $6 + 5 = \boxed{}$

    (c) $4 + 8 = \boxed{}$ (d) $7 + 7 = \boxed{}$

    (e) $9 + 6 = \boxed{}$ (f) $5 + 7 = \boxed{}$

    (g) $6 + 7 = \boxed{}$ (h) $9 + 9 = \boxed{}$

    (i) $8 + 6 = \boxed{}$ (j) $7 + 4 = \boxed{}$

    (k) $6 + 5 = \boxed{}$ (l) $8 + 8 = \boxed{}$

    (m) $9 + 4 = \boxed{}$ (n) $5 + 9 = \boxed{}$

2 Find addition facts.

Exercise 4 • page 111

     6-4 Addition Facts to 20

①

$6 + 7 = \boxed{}$

② (a) $9 + 6 = \boxed{}$
      ?   5

(b) $8 + 5 = \boxed{}$
      ?   3

(c) $7 + 6 = \boxed{}$
      3   ?

(d) $3 + 8 = \boxed{}$
      1   ?

(e) $5 + 7 = \boxed{}$
      2   ?

(f) $4 + 9 = \boxed{}$
      ?   1

③ (a) $8 + 6 = 10 + \boxed{}$

(b) $7 + 7 = 10 + \boxed{}$

(c) $2 + 9 = \boxed{} + 10$

(d) $4 + 7 = \boxed{} + 10$

④ (a) $9 + 7 = \boxed{}$

(b) $9 + 9 = \boxed{}$

(c) $5 + 7 = \boxed{}$

(d) $3 + 9 = \boxed{}$

5 Write an equation for each and find the answer.

(a) Diego put 4 yellow roses in a vase.
Then he puts 9 red roses in the vase.
How many roses are in the vase?

(b) Kona put 8 flowers in a vase.
She has 5 flowers left over.
How many flowers did she have at the start?

6 Put in order from least to greatest.

(a)
| 8 + 5 | 6 + 6 | 10 + 6 | 5 + 4 | 14 + 3 |

(b)
| 15 − 5 | 11 + 4 | 6 + 3 | 17 − 10 | 8 + 4 |

7 Which equations are true?
Which are false?

(a) 9 + 4 = 13

(b) 6 + 8 = 14

(c) 17 = 7 + 7

(d) 13 + 3 = 8 + 8

(e) 8 + 7 = 7 + 9

(f) 7 + 6 = 9 + 4

(g) 7 + 8 = 19 − 4

(h) 18 − 1 = 9 + 9

Exercise 5 • page 113

6-5 Practice

Chapter 7

Subtraction Within 20

Think

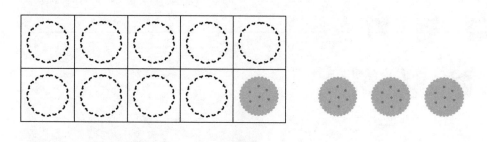

There are 13 crackers.

Alex eats 9 crackers.

How many crackers are left?

Learn

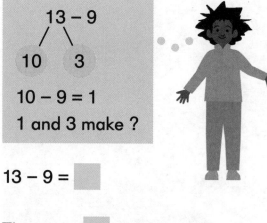

13 − 9

10   3

10 − 9 = 1

1 and 3 make ?

13 − 9 = ▢

There are ▢ crackers left.

<u>Do</u>

1

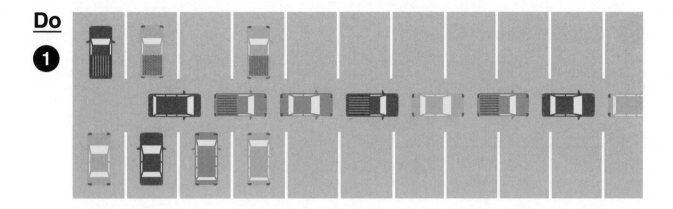

There are 15 cars in the parking lot.

8 cars drive away.

How many cars are left?

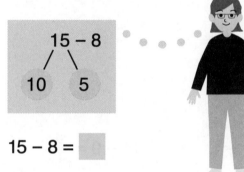

$$15 - 8$$

10   5

$15 - 8 = \boxed{}$

There are $\boxed{}$ cars left.

2 Subtract 8 from 14.

$$14 - 8$$

10   4

$14 - 8 = \boxed{}$

3 Subtract 9 from 15.

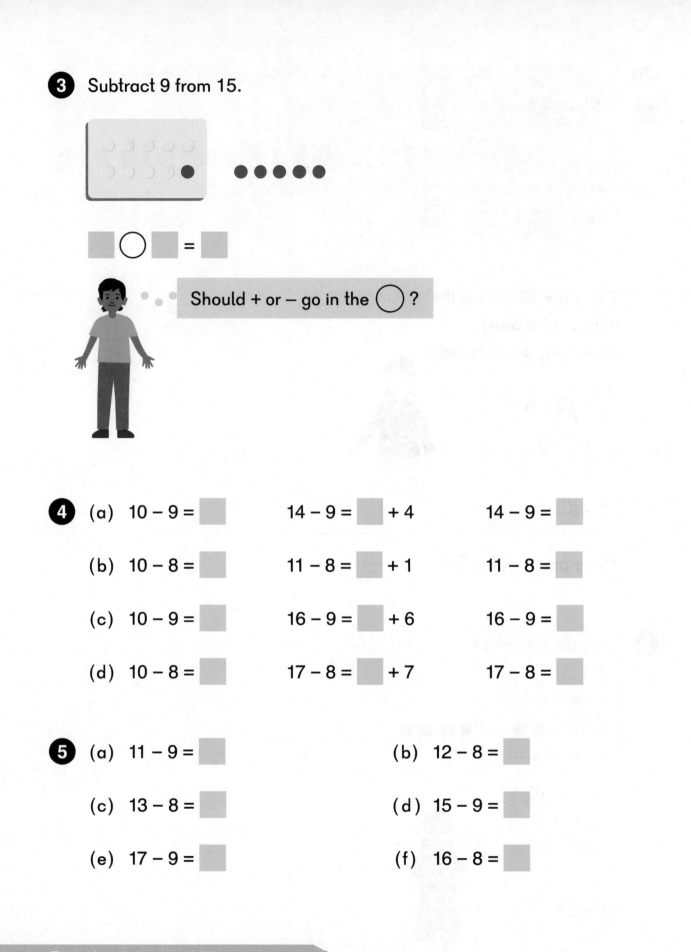

Should + or − go in the ⃝ ?

4 (a) 10 − 9 = ▢   14 − 9 = ▢ + 4   14 − 9 = ▢

   (b) 10 − 8 = ▢   11 − 8 = ▢ + 1   11 − 8 = ▢

   (c) 10 − 9 = ▢   16 − 9 = ▢ + 6   16 − 9 = ▢

   (d) 10 − 8 = ▢   17 − 8 = ▢ + 7   17 − 8 = ▢

5 (a) 11 − 9 = ▢   (b) 12 − 8 = ▢

   (c) 13 − 8 = ▢   (d) 15 − 9 = ▢

   (e) 17 − 9 = ▢   (f) 16 − 8 = ▢

Exercise 1 • page 117

Think

Mei caught 12 trout.

She puts 7 trout back.

How many trout are left?

Learn

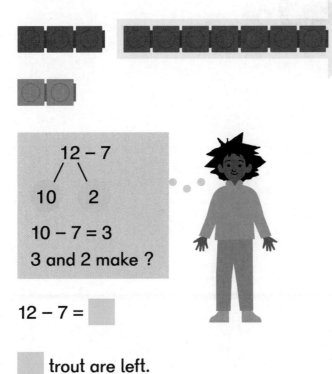

12 − 7

10   2

10 − 7 = 3

3 and 2 make ?

12 − 7 = 🟦

🟦 trout are left.

1 Sofia has 15 stickers.
She uses 6 stickers.
How many stickers are left?

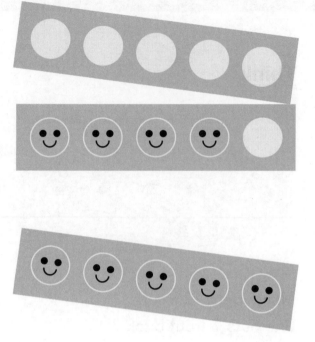

```
   15 – 6
   /    \
 10      5
```

15 – 6 = ▢

There are ▢ stickers left.

2 Subtract 6 from 12.

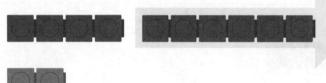

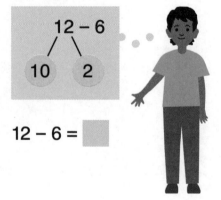

```
   12 – 6
   /    \
 10      2
```

12 – 6 = ▢

3 Subtract 5 from 11.

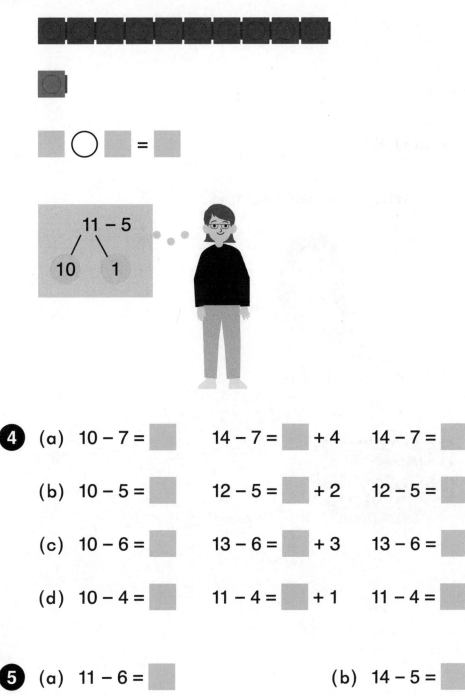

$$\square \bigcirc \square = \square$$

11 – 5

10   1

4 (a) $10 - 7 = \square$ $14 - 7 = \square + 4$ $14 - 7 = \square$

(b) $10 - 5 = \square$ $12 - 5 = \square + 2$ $12 - 5 = \square$

(c) $10 - 6 = \square$ $13 - 6 = \square + 3$ $13 - 6 = \square$

(d) $10 - 4 = \square$ $11 - 4 = \square + 1$ $11 - 4 = \square$

5 (a) $11 - 6 = \square$ (b) $14 - 5 = \square$

(c) $16 - 7 = \square$ (d) $12 - 3 = \square$

(e) $13 - 5 = \square$ (f) $11 - 7 = \square$

Exercise 2 • page 121

Think

Sofia has 12 raspberries.
She eats 5 of them.
How many raspberries are left?

Which ones will I eat first?

<u>Learn</u>

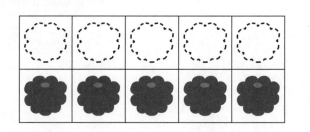

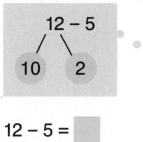

$12 - 5 =$ ⬜

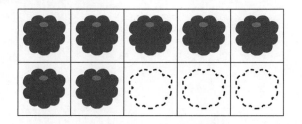

$12 - 5 =$ ⬜

There are ⬜ raspberries left.

Do

1 Subtract 7 from 13.

Method 1

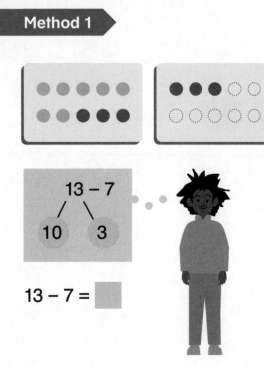

13 − 7

10   3

13 − 7 = ☐

Method 2

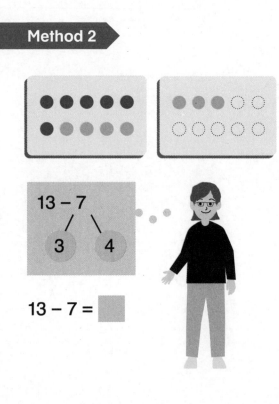

13 − 7

3   4

13 − 7 = ☐

7-3 Subtract the Ones First

2 Subtract 6 from 15.

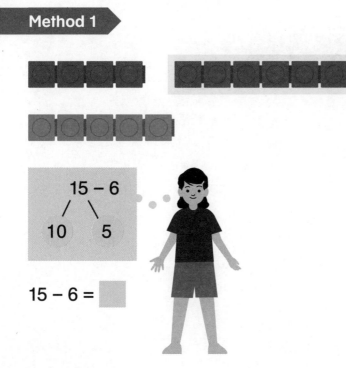

15 − 6

10   5

15 − 6 = ☐

Method 2

15 − 6

5   1

15 − 6 = ☐

3 Subtract 7 from 12.

Method 1

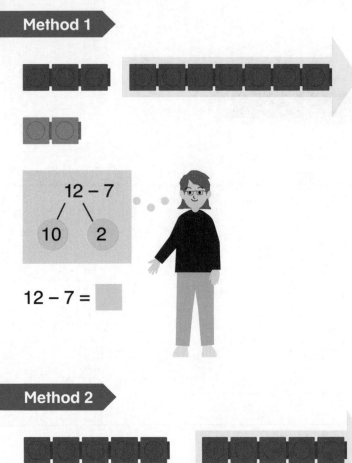

$12 - 7$

10   2

$12 - 7 = $ ☐

Method 2

$12 - 7$

2   5

$12 - 7 = $ ☐

4 Subtract 8 from 14.

Method 1

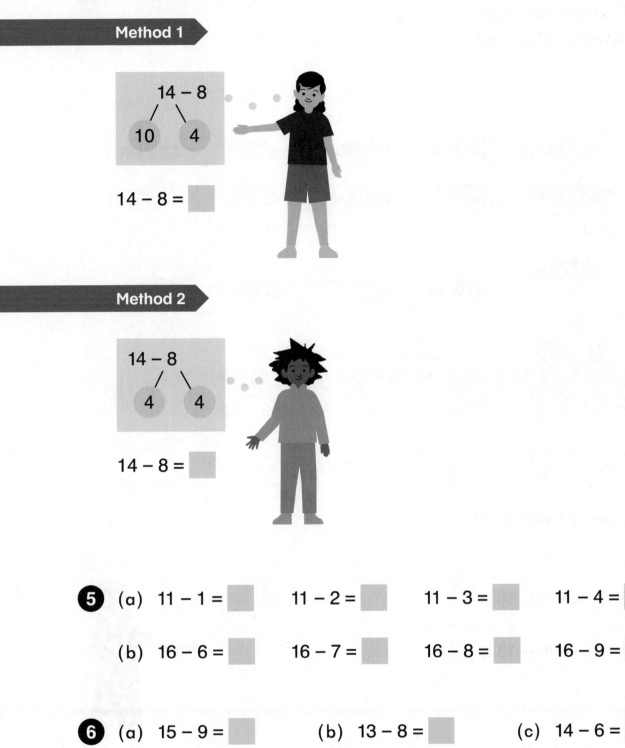

14 − 8

10   4

14 − 8 = ▢

Method 2

14 − 8

4   4

14 − 8 = ▢

5 (a) 11 − 1 = ▢   11 − 2 = ▢   11 − 3 = ▢   11 − 4 = ▢

    (b) 16 − 6 = ▢   16 − 7 = ▢   16 − 8 = ▢   16 − 9 = ▢

6 (a) 15 − 9 = ▢   (b) 13 − 8 = ▢   (c) 14 − 6 = ▢

    (d) 12 − 6 = ▢   (e) 15 − 6 = ▢   (f) 11 − 7 = ▢

7 There are 15 salmon eating.
If 7 salmon swim away,
how many salmon will be left?

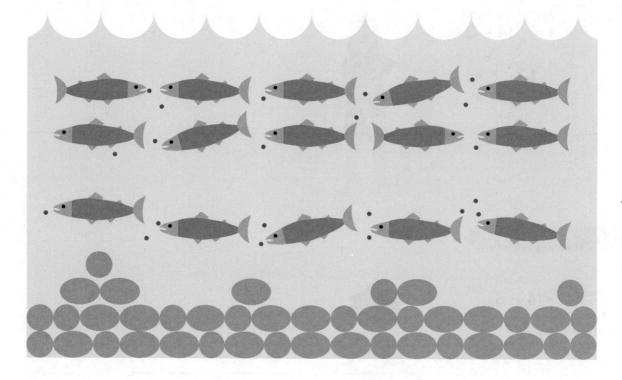

 – ⬜ = ⬜

⬜ salmon will be left.

8 (a) 7 less than 16 is ⬜.

16 – 7 = ?

(b) 6 less than 13 is ⬜.

Exercise 3 • page 125

7-3 Subtract the Ones First

Lesson 4
Word Problems

Think

Write an equation for each story.

There are 11 strawberries.
5 strawberries are on a plate.
The rest are in a bowl.
How many strawberries
are in the bowl?

5 strawberries are on the plate.
6 strawberries are in the bowl.
How many strawberries
are there in all?

Learn

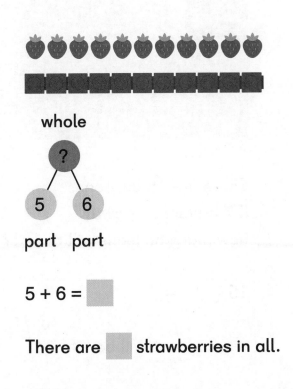

whole

11

5 ?

part part

11 − 5 =

 strawberries are in the bowl.

whole

?

5 6

part part

5 + 6 =

There are ▢ strawberries in all.

Do

1

8 birds are in a bird bath.

If 4 more birds come to the bird bath,

how many birds will be in the bird bath?

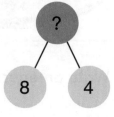

8 ◯ 4 = ▢

▢ birds will be in the bird bath.

2

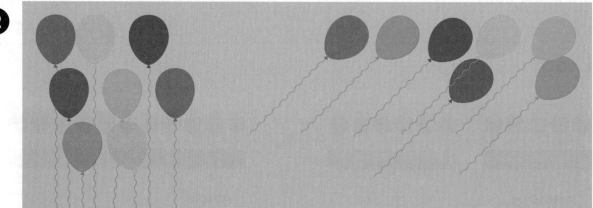

There are 15 balloons.

If 7 balloons fly away,

how many balloons will be left?

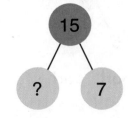

15 ◯ 7 = ▢

▢ balloons will be left.

3

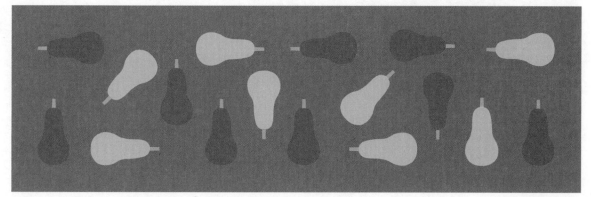

There are 17 squash.
8 squash are yellow.
How many squash are green?

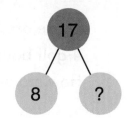

 = ☐

☐ are green.

4

There are 16 bees.
9 bees are on a flower and the rest are on the ground.
How many bees are on the ground?

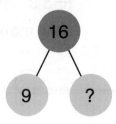

 = ☐

☐ are on the ground.

5

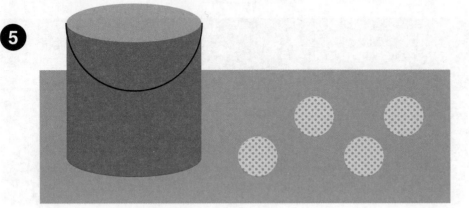

There are 13 golf balls.

4 golf balls are on the grass.

How many golf balls are in the bucket?

13 ◯ 4 = ▢

▢ golf balls are in the bucket.

6

There are 9 nails in the bag
and 8 nails in the box.

How many nails are there altogether?

▢ ◯ ▢ = ▢

There are ▢ nails.

Exercise 4 • page 129

7-4 Word Problems

Think

Say the answers and look for patterns.

Make flash cards for the facts you need to practice.

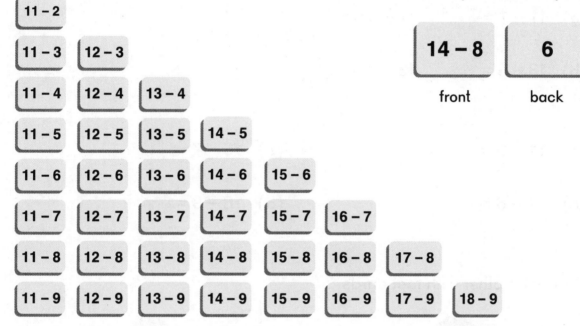

| 11 – 2 |
11 – 3	12 – 3
11 – 4	12 – 4	13 – 4
11 – 5	12 – 5	13 – 5	14 – 5
11 – 6	12 – 6	13 – 6	14 – 6	15 – 6
11 – 7	12 – 7	13 – 7	14 – 7	15 – 7	16 – 7
11 – 8	12 – 8	13 – 8	14 – 8	15 – 8	16 – 8	17 – 8
11 – 9	12 – 9	13 – 9	14 – 9	15 – 9	16 – 9	17 – 9	18 – 9

14 – 8   **6**

front   back

Learn

Put each subtraction fact under the correct answer.

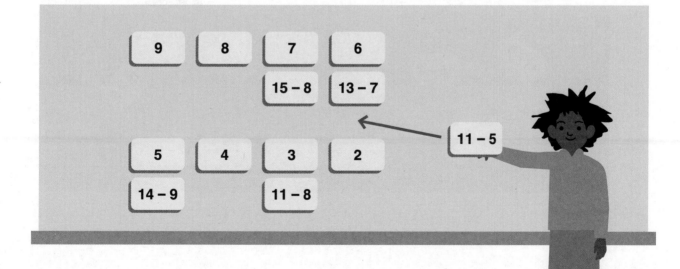

Do

1   (a)   10 − 9 = ☐   14 − 9 = ☐

      (b)   10 − 7 = ☐   13 − 7 = ☐

2   (a)   11 − 1 = ☐   11 − 6 = ☐

      (b)   13 − 3 = ☐   13 − 4 = ☐

3   (a)   11 − 8 = ☐   (b)   16 − 7 = ☐

      (c)   14 − 6 = ☐   (d)   18 − 9 = ☐

4   Quiz each other with fact cards.

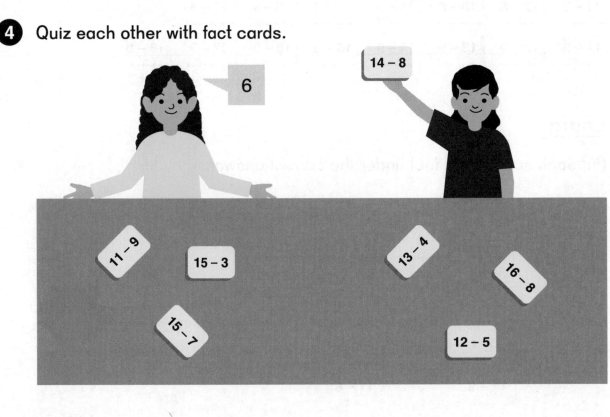

1 (a) $17 - 9 = \boxed{} + 7$

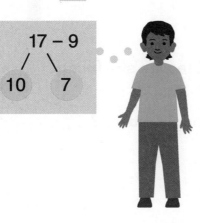

(b) $14 - 8 = 10 - \boxed{}$

2 (a) $15 - 8 = \boxed{} + 5$

(b) $12 - 5 = \boxed{} + 2$

(c) $15 - 8 = 10 - \boxed{}$

(d) $12 - 5 = 10 - \boxed{}$

3 (a) $16 - 7 = \boxed{}$

(b) $14 - 7 = \boxed{}$

(c) $17 - 8 = \boxed{}$

(d) $13 - 6 = \boxed{}$

(e) $18 - 8 = \boxed{}$

(f) $12 - 7 = \boxed{}$

(g) $15 - 9 = \boxed{}$

(h) $11 - 2 = \boxed{}$

(i) $14 - 5 = \boxed{}$

(j) $12 - 8 = \boxed{}$

(k) $13 - 5 = \boxed{}$

(l) $11 - 9 = \boxed{}$

4 Write an equation for each.

(a) 15 birds were sitting on a telephone wire.
6 birds flew away.
How many birds are left?

(b) Lily made a picture with 12 fall leaves she found.
7 of them are maple leaves.
How many fall leaves are not maple leaves?

5 Put in order from least to greatest.

(a) | 14 − 6 | 12 − 7 | 13 − 9 | 15 − 6 | 16 − 9 |

(b) | 13 − 8 | 19 − 3 | 12 − 6 | 18 − 6 | 11 − 8 |

6 Which equations are true?
Which are false?

(a) 11 − 8 = 8   (b) 8 = 13 − 5

(c) 14 − 9 = 1 + 9   (d) 12 + 5 = 15 − 2

(e) 13 − 8 = 10 − 8   (f) 13 − 7 = 3 + 3

Exercise 6 • page 135

7-6 Practice

Chapter 8

Shapes

What did they make?
What shapes did they use?

Think

These are solid shapes.

| Cuboid | Cube | Cylinder | Cone | Pyramid |

What pictures can you draw by tracing around the edges of shapes?

I drew a dog!

Learn

These are **rectangles**.

What do the corners of rectangles look like?

Rectangles have 4 corners.
Rectangles have [ ] sides.

These rectangles are called **squares**.

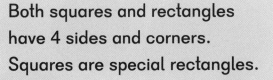

Both squares and rectangles have 4 sides and corners. Squares are special rectangles.

In a square, all [ ] sides are the same length.

These are **triangles**.

The corners on triangles can look different from each other.

A triangle has ⬜ straight sides and ⬜ corners.

These are **circles**.

How are circles different from the other shapes?

A circle has ⬜ straight sides and ⬜ corners.

Do

1 Find rectangles, squares, triangles, and circles on objects around you.

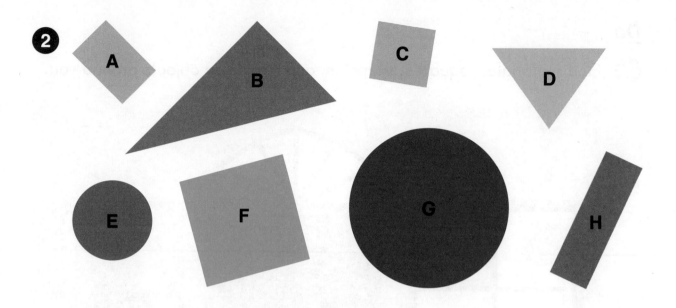

2

(a) Which shapes are rectangles?

(b) Which shapes are squares?

(c) Which shapes are circles?

3 How are the shapes below similar to or different from rectangles?

The corners on these shapes do not look like the corners on rectangles.

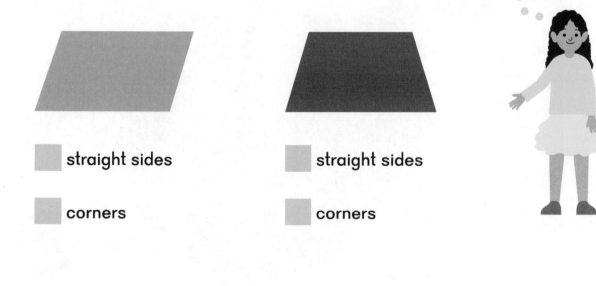

straight sides

corners

straight sides

corners

Exercise 1 • page 139

Think

Group these shapes in different ways.

<u>Learn</u>

How are the shapes grouped?

(a)

(b)

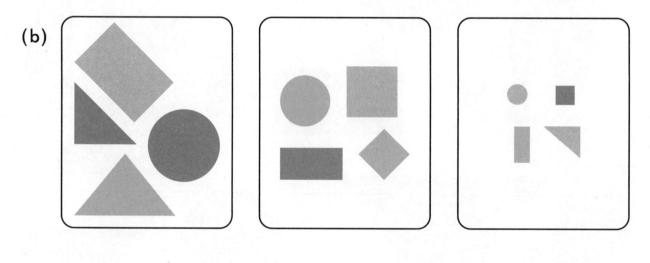

(c)

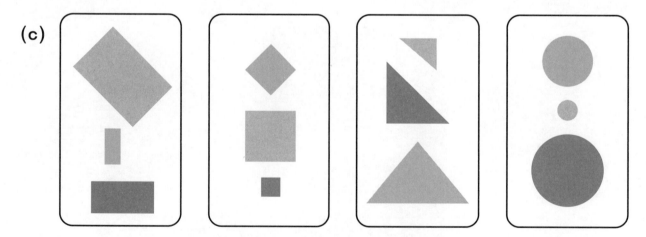

Do

1 How are these shapes grouped?

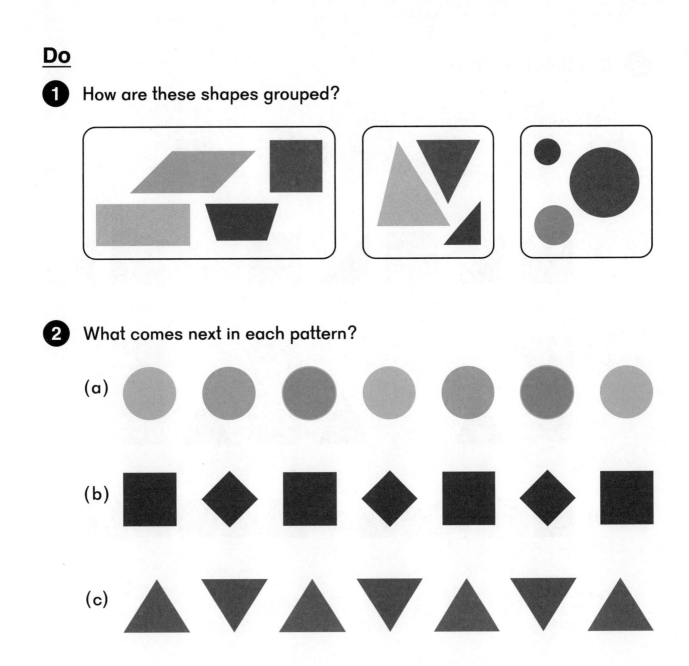

2 What comes next in each pattern?

(a)

(b)

(c)

3 Complete the pattern.

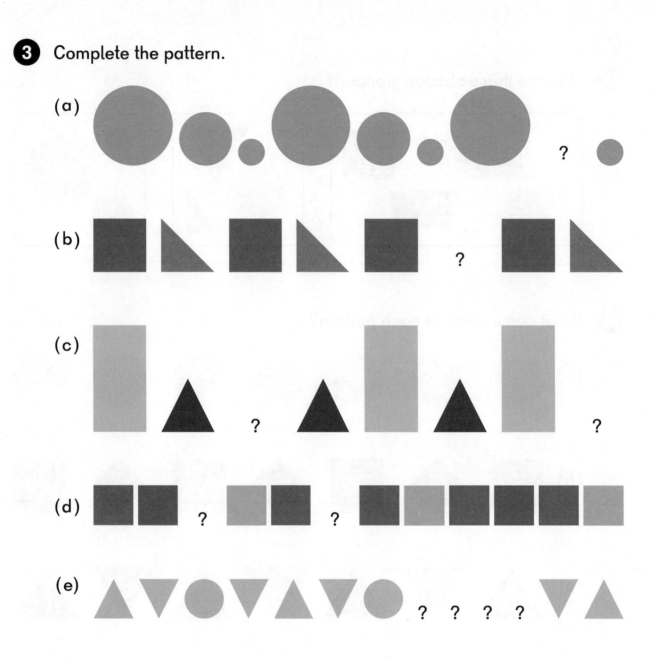

Think

Make pictures using triangles.

Learn

What new shapes are made from these triangles?

Do

1 Put 2 of the triangles together to make these shapes.

2 Put 4 of the triangles together to make these shapes.

3 Put triangles together to make these shapes.

4 Put these shapes together to make circles.

(a)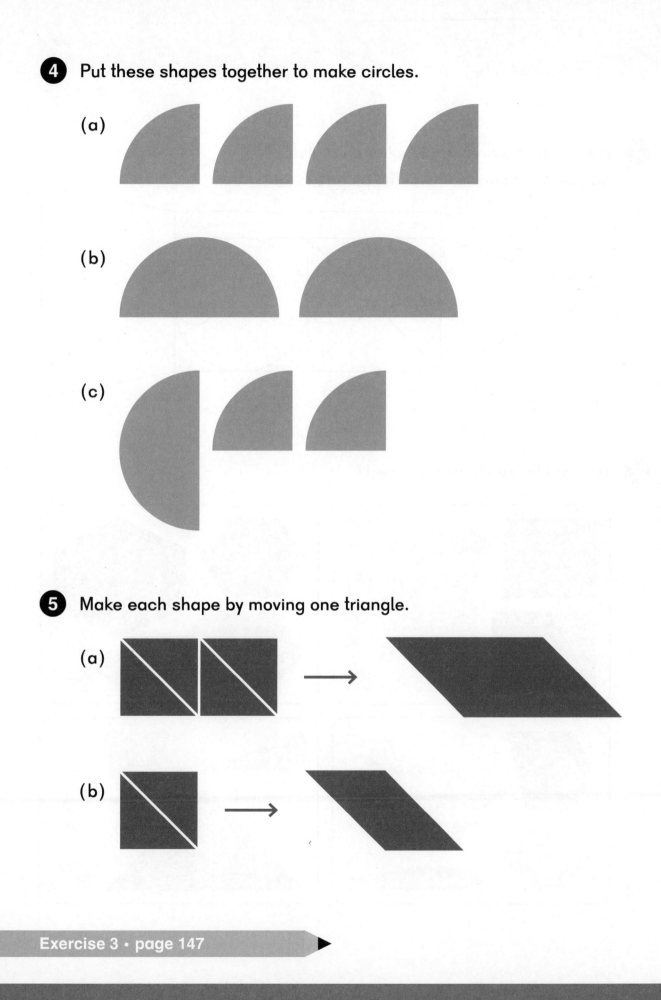

(b)

(c)

5 Make each shape by moving one triangle.

(a)

(b)

Exercise 3 • page 147

1 How many squares, rectangles, triangles, and circles can you find in the picture?

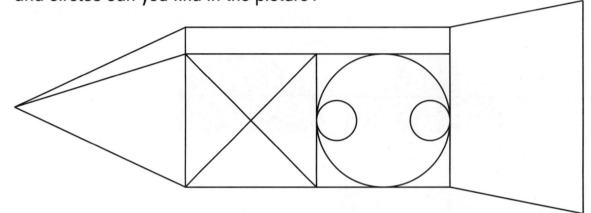

2 How are these shapes grouped?

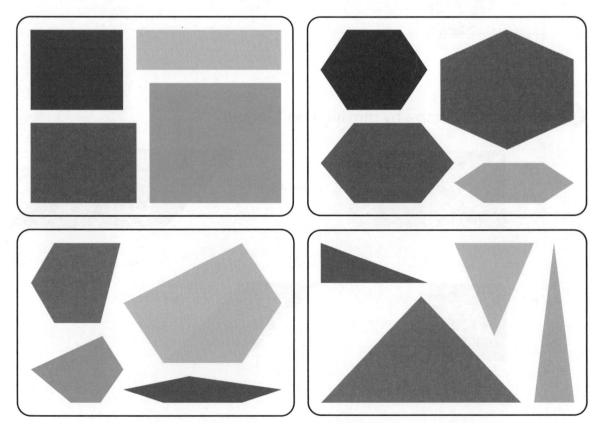

3 Make each shape using these shapes: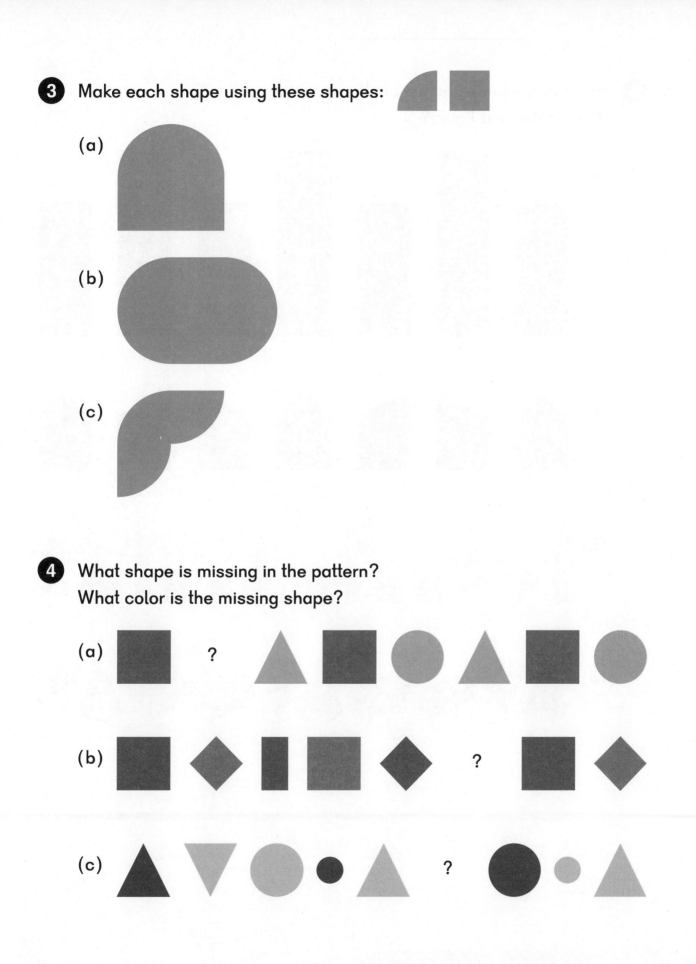

(a)

(b)

(c)

4 What shape is missing in the pattern?
What color is the missing shape?

(a)

(b)

(c)

5 How many figures are there in the repeating pattern?
What is the next figure?

(a)

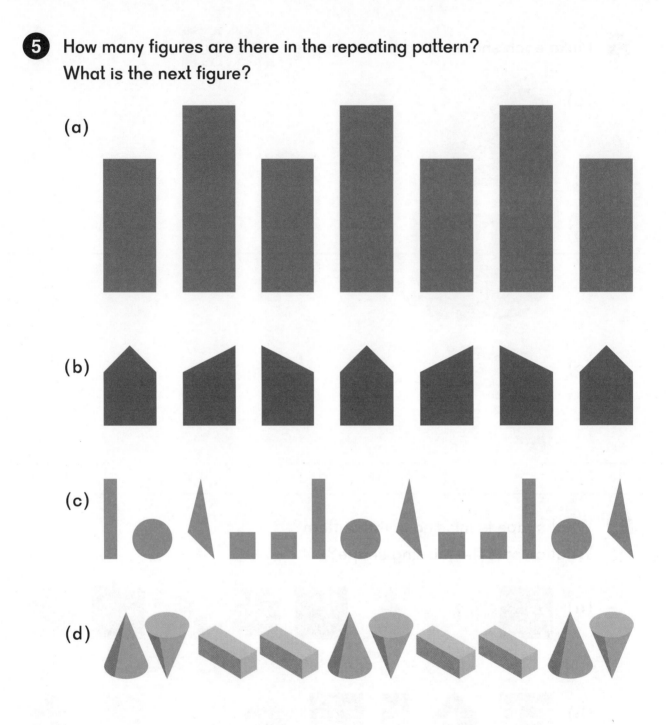

(b)

(c)

(d)

Exercise 4 • page 151

8-4 Practice

Chapter 9

Ordinal Numbers

Think

Who are the **first three people** to cross the finish line?
Who came in **third** place?

Learn

The third place

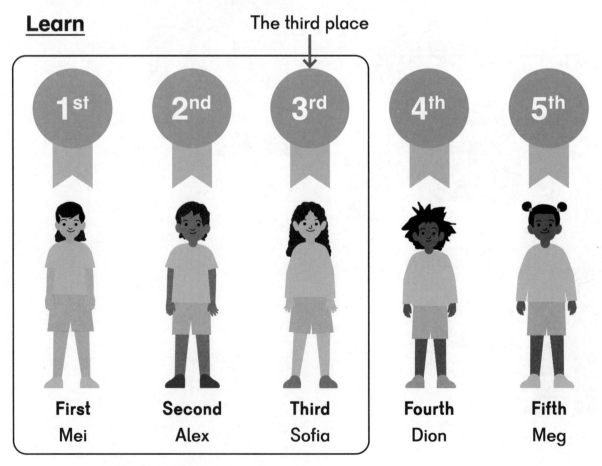

1st	2nd	3rd	4th	5th
First	Second	Third	Fourth	Fifth
Mei	Alex	Sofia	Dion	Meg

The first three people

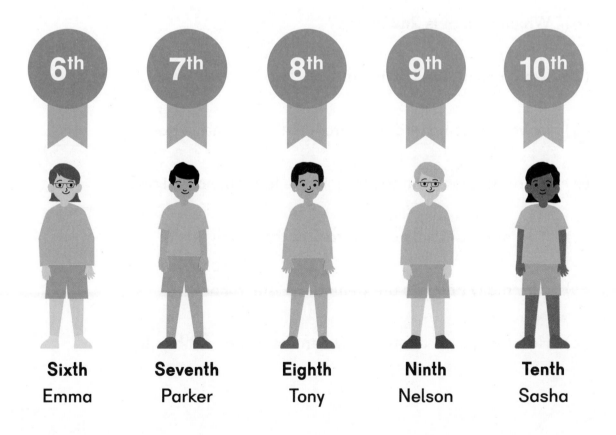

| **Sixth** | **Seventh** | **Eighth** | **Ninth** | **Tenth** |
| Emma | Parker | Tony | Nelson | Sasha |

(a) Which animal is **1st** in line?

(b) Which animals are the **first 5** in line?

(c) Which animal is **5th** in line?

(d) Which animal is **5th from last**?

(e) Which animal is **2nd** in line?

(f) Which animal is **7th** in line?

(g) In which position from the front is the mountain goat?

(h) What position from the front is the last animal in line?

(i) How many animals are **in front of the 8th** in line?

(j) How many animals are **behind the 8th** in line?

(k) How many animals are **between the 6th and 9th** in line?

left ⬤⬤⬤⬤⬤⬤⬤⬤⬤ right

2 (a) How many colors are there?

(b) What color is **3rd from the right**?

(c) What color is **5th from the right**?

(d) What color is **5th from the left**?

(e) In what position from the left is orange?

(f) In what position from the right is orange?

3 (a) How many shapes are there?

top

(b) What shape is **3rd from the top**?

(c) What shape is **3rd from the bottom**?

(d) In what position from the top is the circle?

(e) In what position from the bottom is the circle?

(f) How many shapes are **below the 1st** shape from the top?

bottom

Exercise 1 • page 155

Think

Mei is the 5th child in line.

There are 4 other children after her.

How many children are in line?

Learn

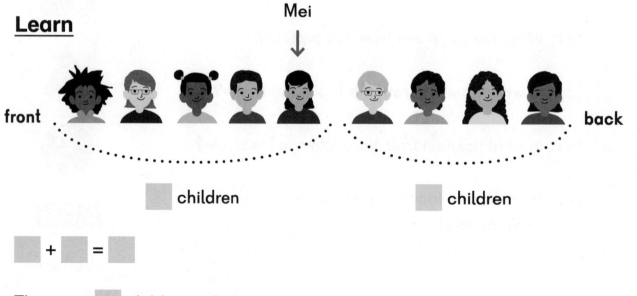

Mei

front .... back

■ children   ■ children

■ + ■ = ■

There are ■ children in line.

Do

 1 There are 15 cars waiting in line.
The blue car is 6th in line.
How many cars are after the blue car?

 2 Sofia is the 4th person from the front of the line
and the 3rd person from the back of the line.
How many people are in the line?

Sofia

 3 Alex is digging holes.
He is now digging his 3rd hole.
He has 5 more holes to dig.
How many holes does he have to dig in all?

 4 There are 9 people standing in line.
Emma is the 4th person in line from the front.
Dion is the 3rd person from the back of the line.
How many people are standing between them?

Exercise 2 • page 159

left right

❶ (a) Which fruit is 7th from the left?

(b) In what position from the right is the orange?

(c) In which position from the left is the pear?

(d) How many fruit are between the 1st from the left and the 2nd from the right?

❷ Mei's drawing is 4th from the top and 3rd from the bottom. How many drawings are there?

3 Connor is 6th in line.
There are 8 more people behind him.
How many people are in the line?

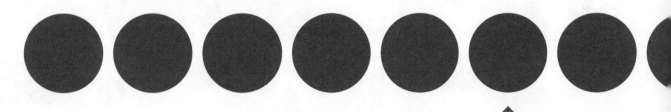

Connor

4 Dana is in a line.
There are 5 children in front of her and 4 children behind her.
How many children are in the line?

5 Ava is in a line.
She is 5th from the front and 4th from the back.
How many children are in the line?

6 Jack is in the middle of a line.
There are 7 children in the line.
What is Jack's position from the front?

Exercise 3 • page 163

1

(a) There are ▢ rectangles, ▢ triangles, and ▢ circles.

(b) There are fewer _____ than triangles.

(c) The number of _____ is the greatest.

(d) Add 6 more circles. There are now ▢ circles.
Write an equation.

(e) Take away 8 triangles. There are now ▢ triangles.
Write an equation.

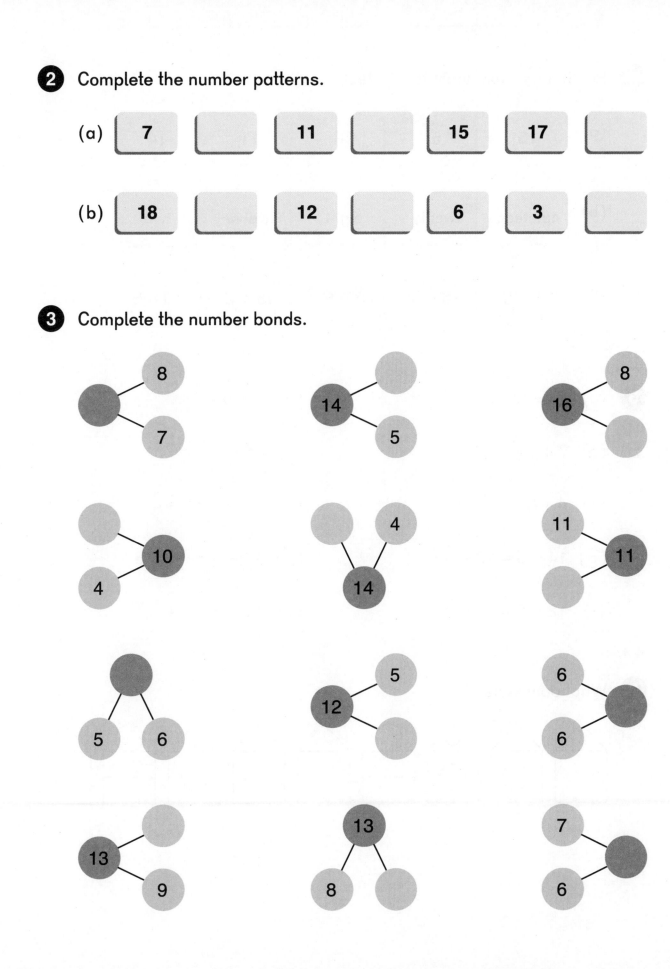

2 Complete the number patterns.

(a) | 7 | | 11 | | 15 | 17 | |

(b) | 18 | | 12 | | 6 | 3 | |

3 Complete the number bonds.

4 Put in order from least to greatest.

(a)

| 15 | 6 | 18 | 11 | 14 |

(b)

| nineteen | twenty | eight | twelve | two |

(c)

| 15 − 6 | 16 − 6 | 9 + 6 | 15 + 2 | 19 − 6 |

5 (a) $14 - 6 = \boxed{}$ (b) $8 + 9 = \boxed{}$

(c) $7 + \boxed{} = 11$ (d) $15 - \boxed{} = 7$

(e) $4 + 8 = 10 + \boxed{}$ (f) $16 - 7 = 3 + \boxed{}$

(g) $14 - 5 = 10 - \boxed{}$ (h) $13 - 8 = \boxed{} - 5$

(i) $12 - 8 = 13 - \boxed{}$ (j) $11 - 2 = 4 + \boxed{}$

6 What comes next?

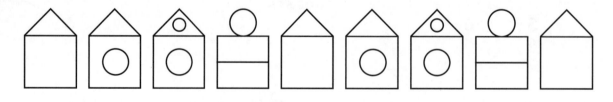

Write an equation for each and find the answer.

(a) Mari has 7 books.
She buys 5 more books.
How many books does she have now?

(b) Salim wants to read 15 books.
He has read 9 books so far.
How many more books does he still need to read?

(c) Tyler found 15 seashells.
3 of them are broken.
How many seashells are not broken?

(d) Holly gave away 3 seashells.
She now has 6 seashells.
How many seashells did she have at first?

(e) There are 14 dogs in an animal shelter.
7 of the dogs are adopted.
How many dogs are waiting to be adopted?

(f) 4 cats were adopted in the morning.
8 cats were adopted in the afternoon.
How many cats were adopted that day?

8 Lisa read from the beginning of the 3rd page
to the end of the 12th page in a book.
How many pages did she read?

9 These figures are grouped into 4 rows and 7 columns.

top

left   right

bottom

(a) How are they grouped into rows?

(b) How are they grouped into columns?

(c) What are the shapes and colors of the missing figures?

(d) What color is in the middle column?

(e) What shapes are in the 3rd row from the bottom?

(f) What color is in the 3rd column from the right?

(g) What shape and color is 4th from the left and 4th from the top?

(h) In what position is the green triangle?

Exercise 4 • page 167